First published in the United Kingdom in 2012 by
Portico Books
10 Southcombe Street
London
W14 0RA

An imprint of Anova Books Company Ltd

ISBN 9781907554698

A CIP catalogue record for this book is available from
the British Library.

10 9 8 7 6 5 4

Printed and bound by G. Canale & C.SpA, Italy

This book can be ordered direct from the publisher at
www.anovabooks.com

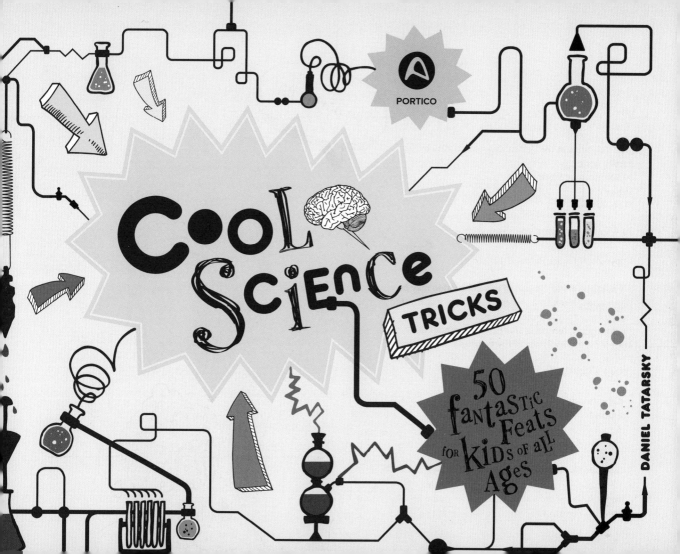

CONTENTS

Introduction

At first glance the title of this book appears to be an oxymoron: how can science be cool? When most of us think of science, we can't help but remember the interminably dull science lessons we endured at school (I loved physics but I was alone amongst my friends on this one). So it's not surprising that we would be unlikely to put the words 'cool' and 'science' in the same sentence, let alone into a book!

The funny thing is that when you do think back to your third- and fourth-year chemistry lessons what comes to mind is often really exciting: who can forget the first time you made your own sparklers out of iron filings? Or the day you took your best mate's blood and worked out their blood group? Or how about that morning when Mr Baines brought in a pig with two heads – that was definitely not on the curriculum!

I digress, but the point is that science can be surprising, exciting and yes, possibly very cool. This book includes 50 examples of such coolness and the best thing is that you can do all of it in the safety of your own home (well, almost all of it!), and these experiments can be used to impress kids and adults alike. Jaws will hit the floor when you unveil the balancing coke can trick, gasps will fill the air when you whip a tablecloth off the table leaving everything on it exactly where it was, and eyes will be rubbed in disbelief when you explain how to get a boiled egg into a milk bottle.

With not much practice and just a few props, you will become the coolest dude (or dudette) in the room. So without further ado, let's dig out that Bunsen burner, dust off the thermometer and put on our goggles … it's showtime!

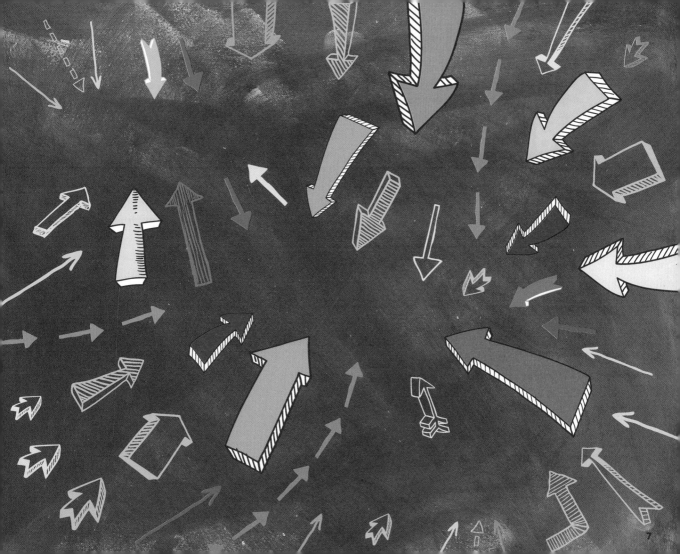

Before we begin ...

The tricks detailed in this book have different safety provisions. There are some tricks that will require parental supervision. Please make sure they are present and, most importantly, paying attention!

Some tricks require no additional safety measures but please be mindful at all times when carrying out any of the tricks – you don't want to poke an eye out or make a mess. So, be careful, enjoy the tricks – and learn some cool science!

SAFETY NOTICE

FOR EVERYONE

Tricks that everyone can do, any time. Go for it!

PLEASE BE CAREFUL

Please read the instructions carefully before beginning. If you are under the age of 16, please make sure an adult is with you and keeping an eye on what's going on.

DANGER – ADULTS ONLY

Even though adults can be just as clumsy as children (sometimes even more so!), please make sure that an adult sets up and completes this trick and that all safety checks are carried out beforehand. And if they make a mess, make sure they clear it up!

The Phantom Coin

FOR EVERYONE

Money burning a hole in your pocket is a well-known phenomenon (sadly not one of which I have much experience), but how about burning a hole in your forehead? It should be easy to dislodge a coin stuck to your head, but this trick shows that it can be pretty much impossible.

Stuff To Get
- A coin
- A willing volunteer

The Science

Our skin is very sensitive, but the level of sensitivity varies depending on which part of the body you are touching, because in some areas the nerve endings are more densely packed. For instance, your palms are much more sensitive than the backs of your hands.

This trick relies on the signals sent to the brain when the coin is pressed against it: your volunteer feels the sensation of the coin touching their forehead, and even when you take the coin away, the nerve endings continue to send signals to the brain.

Did You Know?

A similar, but more extreme, phenomenon is phantom limb syndrome, which is experienced by amputees who continue to feel movement or even pain from limbs they no longer have.

Let's Get Started

1. This trick only works if you demonstrate it first, so take a coin and press it onto your forehead. Hold it there for about five seconds and then let go. The moisture in your skin will hold it in place.

2. Your job now is to dislodge it by scrunching up your forehead. Easy – it drops off quite quickly. Once you have shown how easy it is, ask for a volunteer to see if they can do it more quickly.

Make your friend think the coin is still there!

Signals sent to the brain!

Nerve endings at work!

3. Repeat step 1, but, when you have held the coin onto your volunteer's forehead for five seconds, don't let go of it – just make your stooge think you have. What you actually need to do is take it off and pocket it.

4. Now stand back and watch them contort their face while they attempt to dislodge the coin. If they are still trying after ten minutes, it may be kind to explain to them what you have done.

Demonstrate it yourself first – for the full effect

The Effervescent Rocket

PLEASE BE CAREFUL

Men (and women) of science had wanted to explore space for centuries. Then the Space Race happened and it became a reality. When Soviet cosmonaut Yuri Gagarin became the first person to leave the Earth's atmosphere in 1961, it was the brains of rocket scientists that got him there. Although for this awesome trick you don't need to be a rocket scientist ...

Did You Know?

There was a time when Alka-Seltzer was marketed as a panacea – a cure for everything. This did not last long and it is now used for mild pain relief, and various science tricks.

Stuff To Get
- An effervescent pain-relief tablet (Alka-Seltzer or similar)
- A small plastic container with a press-on lid, e.g. an old 35mm camera film container
- Water

Let's Get Started

1. Fill the container half full with water, drop the tablet in and quickly put the lid on tightly. Turn the container upside down and place it, lid down, on a table.

2. Wait.

3. KA-BOOM!

Newton's Third Law of Motion – Zoom!

(NaHCO₃ and (C₆H₈O₇) reaction!

The Science

When water is introduced, the reaction between the sodium bicarbonate ($NaHCO_3$) and citric acid ($C_6H_8O_7$) in the effervescent tablet creates carbon dioxide – the bubbles. Lots of bubbles! The quick build-up of carbon dioxide continues until the pressure inside the container forces the lid and the rest of the container apart – we have blast off!

This trick also demonstrates Newton's Third Law of Motion: all forces in the Universe occur in equal but oppositely directed pairs. The gas produced inside the container is forced out at the weakest point (the lid) in one direction (downwards) while the rest of the container moves in the opposite direction (upwards).

The Bathtub Speedboat

Bathtime doesn't just have to be about getting clean. With this nifty piece of work you can turn your tub into a millionaire's playground. Okay, that's a slight exaggeration, but this trick is still a lot of fun.

FOR EVERYONE

Stuff To Get
- A few drops of washing-up liquid
- An ice lolly stick
- Water
- A bath

Let's Get Started

1. After a hard day at work everyone needs a nice soak in the bath. So, fill your bath up, but don't get in just yet, and DON'T use any bubble bath – I know, it's tempting …

2. When you have a level of water you are happy with, place the ice lolly stick at one end. It won't sink, don't worry. Drip a couple of drops of washing-up liquid on one end of the stick and stand back while it whizzes in the opposite direction.

3. This will have woken you up again so it is now time to take the stick out of the bath and hop in for a lovely relaxing soak. And you can finally add your bubble bath.

The Science

In a bath the water molecules beneath the surface are pulled together equally in all directions, but those on top are pulled together more tightly because they only have molecules to their sides and underneath. This difference in force draws the water molecules at the surface together, forming a 'skin' better known as surface tension, and it is this that stops the stick from sinking. Adding washing-up liquid disrupts the arrangement of the water molecules. The water molecules near the washing-up liquid are attractedto the detergent as well as to other water molecules, so the surface tension of the water behind the stick decreases. Water molecules move from areas of low surface tension to areas of high surface tension. The stick is pulled towards areas of high surface tension by the water in front of the stick.

Step Through A Postcard

PLEASE BE CAREFUL

Many postcards have the phrase 'Wish you were here' emblazoned on the front. These familiar words can be interpreted both as a question or an expression of desire. Either way, this jaw-dropping trick allows you to go where no one has gone before ...

Let's Get Started

1. Fold your postcard in half along its long axis. Make sure it has no sentimental value – you won't be able to return it to its original state afterwards!

2. Make a cut from the fold to within 0.5cm (⅕in) of the edge, about 0.5cm (⅕in) away from one end. Do this again all along the postcard until you reach the end.

3. In between each of the cuts you have made along the folded edge, make cuts from the open edge of the card, again stopping about 0.5cm (⅕in) from the edge.

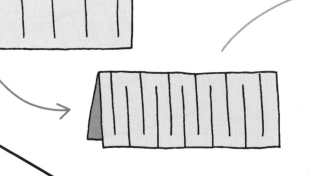

Austrian Dr Emanuel Hermann has the following inscription on his gravestone: 'Der Erfinder Der Postkarte'. He wrote an article in 1869 campaigning for a lower postage rate for postcards and is thus credited by some as the inventor of postcards.

Did You Know?

4. Now cut across the folded edge but leave the two ends. You will now be able to open out the card, and step through the hole in the paper chain you have created.

Wish you were here!

Cut away from the fold

The Science

What is happening here is that you are actually cutting the rectangle of card into one long loop. Seems impossible? Not if you know how!

Bouncy Bouncy

FOR EVERYONE

Everyone, everywhere, wants more bounce in their balls and this will give you supercharged balls that will be the envy of your friends and neighbours. Stop sniggering at the back and let us begin ...

Stuff To Get
- A tennis ball
- A football

Let's Get Started

1. Make a bet with your friends and challenge them to see how high they can make a ball bounce just by dropping it – they will try to hold the ball as high as they can and bounce it as hard as they can to gain extra momentum.

2. After they have tried for a while, move them back to make some room. Hold the football in one hand, and the tennis ball in the other. Before you drop the balls make sure that they are touching one another with the tennis ball directly on top of the football.

3. Let go of the balls; when they land the football will bounce to its normal height but the tennis ball will shoot up way beyond anyone else's effort.

Did You Know?

The official match ball for the 2010 World Cup held in South Africa, the Adidas Jabulani (which means 'celebrate' in Zulu), was described as the 'roundest football ever'. It was made from eight spherically moulded panels and the surface was textured with grooves to improve aerodynamics.

Energy is released

The tennis ball goes w h o o s h !

The Science

Holding a football or tennis ball in the air gives it potential energy. Potential energy can be described as stored energy, and is there due to an object's position (a ball 3m (10ft) off the ground has more potential or stored energy than one 1.5m (5ft) off the ground), or the arrangement of its molecules (a squashed ball has more potential energy than an unsquashed one).

When you drop the ball and it hits the floor, that energy is absorbed into the ball as it is squashed into the floor and is then released when the ball returns to its normal shape, propelling it upwards.

With the double-ball bounce technique in our trick, you get a higher bounce because the tennis ball releases its own potential energy and when this is added to the potential energy of the football – pushing up off the ground at the same time – it means that the tennis ball will go shooting up into the air and you win the bet.

The Möbius Strip Teaser

Nothing lasts forever, or so they say. This trick proves that some things do go on, and on, and on and ... you get the picture.

PLEASE BE CAREFUL

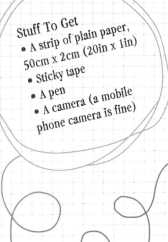

Stuff To Get
- A strip of plain paper, 50cm x 2cm (20in x 1in)
- Sticky tape
- A pen
- A camera (a mobile phone camera is fine)

Let's Get Started

1. Hold each end of the strip of paper between the thumb and forefinger of each hand. Twist one end of the strip once through 180° and hold the two ends together.

2. Ask a friend, a relative or a friendly passer-by to hold the join while you use the sticky tape to secure the ends of the paper together. Now give the pen to your helper/friend and ask them to draw a line down the middle of one side of the strip. Tell them to start at the join and keep going until they return to the starting point.

3. While they are concentrating on drawing a lovely straight line you can get your camera ready. As they finish the line, point out that they have drawn on both sides of the paper without lifting the pen up. As they realize this, take a picture of the look of astonishment on their face.

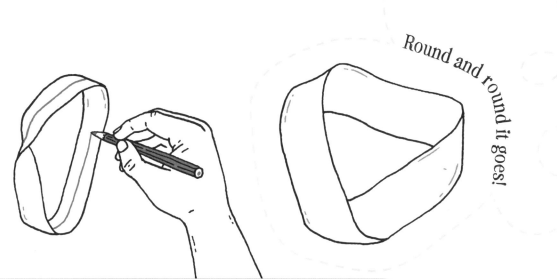

Round and round it goes!

The Science

The Möbius strip was discovered independently by two German mathematicians, August Ferdinand Möbius and Johann Benedict Listing, in 1858. It is a one-sided, non-orientable surface, and, like the cylinder, it is not a true surface but a surface with a boundary.

In mathematics the Möbius strip can be represented using Euclidean (three-dimensional) geometry or topology, which is an area of mathematics concerned with the stretching of different shapes. However, let's keep it simple here, and say that in effect what is happening in this trick is that we are joining the top side with the bottom side so that they become one.

Did You Know?

The Möbius strip has many practical applications in science, technology and industry. One of the most common applied uses is on the magnetic recording tape previously used in the music and radio industry. Rather than using normal tape that can only record on one side, by recording with a Möbius strip, the tape records on both sides – and lasts twice as long.

Super Cool Water

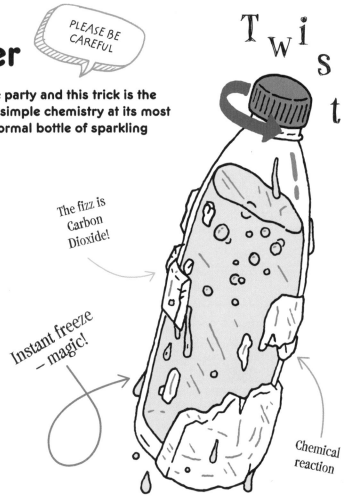

PLEASE BE CAREFUL

Everyone wants to be the coolest one at the party and this trick is the ultimate ice-breaker. This deft little stunt is simple chemistry at its most amazing and centres around a seemingly normal bottle of sparkling water that freezes instantly when opened.

The Science

Fizzy water, because it contains carbon dioxide and a little salt, freezes at a lower temperature than pure water (around -8°C/17.6°F). This means it can go below the normal freezing point of water without freezing. When the bottle is opened, the bubbles are released and the carbon dioxide is expelled (that's the fizz you hear as you unscrew the lid), and the freezing point of the water rises … causing the water to freeze before your eyes.

Twist

The fizz is Carbon Dioxide!

Instant freeze – magic!

Chemical reaction

- A plastic bottle of sparkling water; do NOT use a glass bottle.
- A freezer

Let's Get Started

1. Put a bottle of sparkling water in the freezer. A 500ml (17½ fl oz) bottle will need about two hours, but the timing will vary depending on the temperature of your freezer, and the size of your bottle.

2. If you are hosting a party, you could offer a designated driver – or anyone you think might be keen to see some awesome magic – if they would like a drink of 'magic' sparkling water.

3. Take the bottle out of the freezer, hand it to the intended target, and ask them to open it. As they twist off the lid, the contents will immediately freeze in front of their eyes. Get ready to catch them – they may faint!

Carbonated water was first developed by clergyman and chemist Joseph Priestley. In 1767 he moved next door to a brewery in Leeds and started to experiment with a brewery gas known as fixed air. In 1772 he announced his invention of soda-water, which was produced by impregnating water with fixed air. It was believed the drink might prevent scurvy.

Did You Know?

Bending Light

PLEASE BE CAREFUL

It is a well-known scientific fact that light travels in straight lines. How better then to impress your chums than defying physics and making light bend. As Yoda might say: 'Having a light sabre, it is almost as cool as'.

The Science

Light does indeed travel in a straight line along the path of least resistance. It is one of the constants of the Universe.

In this trick, as the light goes through the water and reaches the point where the stream of water hits the air, it reflects off that boundary and back into the stream. Each time it hits the edge it is reflected back, and so appears to curve. It is this property of light that enables it to travel along optical cables – which is great because if it didn't you'd be reading this in the dark!

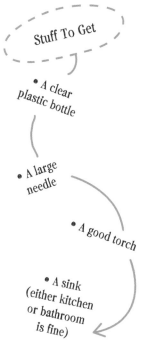

Stuff To Get

• A clear plastic bottle

• A large needle

• A good torch

• A sink (either kitchen or bathroom is fine)

Let's Get Started

1. Announce to your friends that science is rubbish and you can disprove it. They will scoff at your claim but will (hopefully) gather round out of curiosity.

2. Fill the bottle with water and put the lid back on. Carefully punch a small hole in the bottle about a quarter of the way up from the bottom using a needle; because the hole is small the water should not squirt out. Place the bottle next to

... et voila!

Watch the light curve

Torch shines through bottle

the sink with the hole pointing towards the sink.

3. Turn out the lights. Shine the torch through the bottle directly opposite the hole. Take the lid off the bottle.

4. As the water squirts out of the hole it will be illuminated by the torch, and the light will go down the flow of water into the sink: you have defied physics and made light bend. Well done!

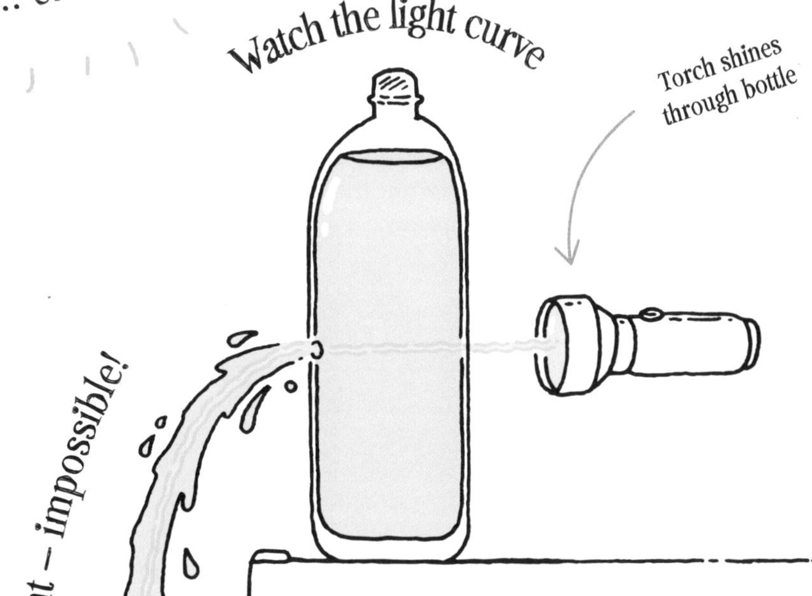

Bendy light – impossible!

Did You Know?

Born in AD 965, medieval Islamic scientist al-Hassan Ibn al-Haytham conducted early investigations into light. Among many other things, he proved that we are able to see because light enters our eyes, rather than the prevailing wisdom of the time, favoured by Plato, Euclid and Ptelomy, that light shines from our eyes onto the objects we see.

Can Crusher

Action heroes and strong men often crush cans to demonstrate just how tough they are. But you too can be a tough guy and outsmart your chums by crushing a can ... with science! Yep, believe it or not, your brain can defeat their brawn. Remember that the next time someone steals your lunch money.

DANGER – ADULTS ONLY

Water pressure will act the same way as pressure in the air. If you put a sealed empty can on the surface of the ocean it would float and be fine. However, if you were able to take it to the bottom of the ocean, it would be crushed by the water pressure – and so would you!

Did You Know?

The Science

Before it is heated, the can is filled with water and air. When the water is boiled it changes state from a liquid to a gas (water vapour). The water vapour pushes the air that was originally inside the can out into the atmosphere. When you turn the can upside down and place it in the water, the water vapour condenses and changes back into water. In its liquid state water molecules are many times closer together than in its gaseous state: the water vapour that filled up the inside of the can turns into only a drop or two of liquid, which takes up much less space.

This small amount of water is unable to exert much pressure on the inside walls of the can and so the pressure of the air pushing from the outside of the can is great enough to crush it.

Let's Get Started

1. Ask a friend to crush a can – this will make them feel tough. Now tell them that you can do it without using your hands – they will laugh and think you are stupid. Ignore them and put a tablespoon of water into the second can.

2. Place the can on the hob and turn on the heat. Bring the water up to boiling point. At this point water vapour will start to come out of the can. Allow it to boil for about 30 seconds.

3. Carefully pick up the can using the tongs and the oven gloves and place it upside down in the cold water – an invisible force will crush the can.

Stuff To Get
- Two clean, empty drink cans
- A bowl of cold water
- A pair of kitchen tongs
- Oven gloves
- An oven hob or hot plate

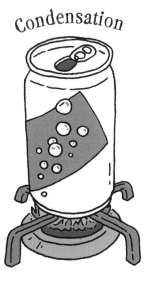

Condensation

Pressure crush!

Cold water

Watch the can shrink

Coin Snatching

PLEASE BE CAREFUL

Money, it's a funny old thing, isn't it? Don't you find it always slipping through your fingers? This awesome trick relies on you being quicker than gravity: yes, it's the classic 'balance a coin on your elbow and then catch it' trick. As we all know, you can't take it with you, so why not put your money to good use ...

Did You Know?
The current world record for coin snatching is 328 coins and was set by Dean Gould from Great Britain in 1993. Mr Gould also holds the records for beer mat snatching and pancake tossing.

The Science
The coin does not hang in the air, of course it doesn't, but it does take a moment more to start moving than your elbow does. And for this we have the principle of inertia to thank. This is covered in Newton's First Law of Motion, which describes an object's resistance to a change in its velocity. The coin, thanks to inertia, resists gravity for long enough for your flying hand to grab it.

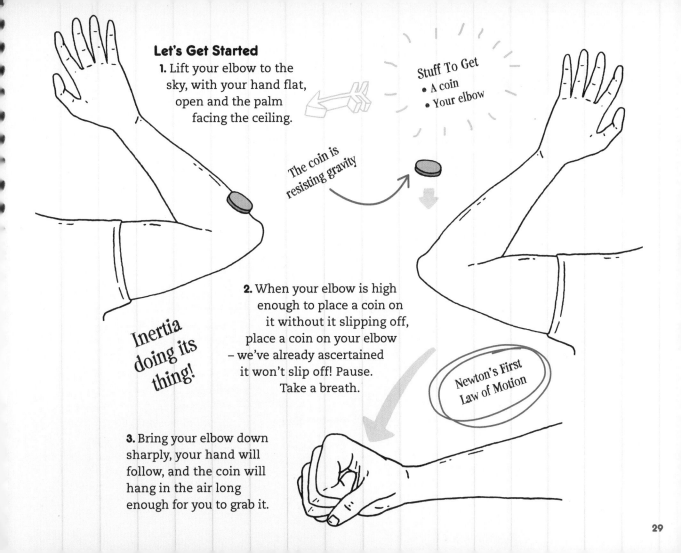

Let's Get Started

1. Lift your elbow to the sky, with your hand flat, open and the palm facing the ceiling.

Stuff To Get
- A coin
- Your elbow

The coin is resisting gravity

2. When your elbow is high enough to place a coin on it without it slipping off, place a coin on your elbow – we've already ascertained it won't slip off! Pause. Take a breath.

Inertia doing its thing!

Newton's First Law of Motion

3. Bring your elbow down sharply, your hand will follow, and the coin will hang in the air long enough for you to grab it.

Hot Flush

Opposites attract, supposedly, but not always. Sometimes it depends on other factors, like science. This experiment demonstrates how heat can affect water and its ability to mix.

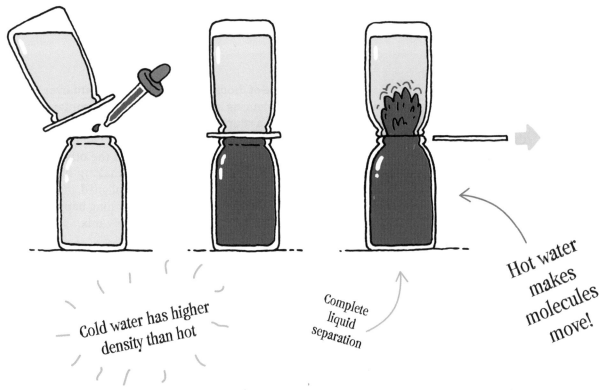

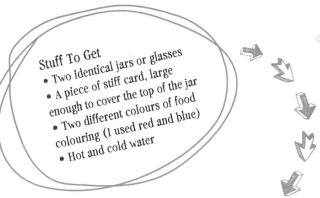

Stuff To Get
- Two identical jars or glasses
- A piece of stiff card, large enough to cover the top of the jar
- Two different colours of food colouring (I used red and blue)
- Hot and cold water

Did You Know?

Most substances are at their most dense when they are frozen, but water is an exception. It is at its most dense at around 4°C (39.2°F), and then as it freezes it becomes less dense once again, and this is why it floats.

Let's Get Started

1. Fill one jar with cold water. Run the hot tap until the water is slightly hotter than you might have it in the bath. Fill the other jar with hot water.

2. Put a couple of drops of red food colouring in the hot water. Put a couple of drops of blue food colouring in the cold water.

3. Hold the card over the top of the jar containing the hot water, turn it upside down and place it directly on top of the other jar. Slowly, making sure the jars are aligned, pull out the card. Nothing happens – they won't mix.

4. Repeat steps 1–3, but this time put the jar containing the cold water on top – holy moly, they're mixing!

The Science

Liquids with a lower density will sit on top of those with a higher density. What this trick proves is that hot water has a lower density than cold. This is because the temperature of the hot water causes the molecules in it to move around, creating larger spaces between them than in cold water, thereby reducing its density.

Balloon on a Skewer

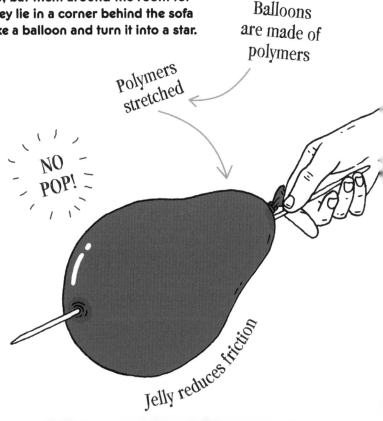

PLEASE BE CAREFUL

Everyone loves balloons: you blow them up, bat them around the room for a while and then forget about them. Then they lie in a corner behind the sofa and slowly deflate. Don't let this happen, take a balloon and turn it into a star.

The Science

All things are made from molecules that are so tiny that we can't see them. Balloons are made up of polymers, which are strings of molecules. Under a microscope you would see the gaps between them.

At the side of a balloon, the polymers are stretched to their tightest, so that, when the tip of a skewer tries to get through, the molecules have no room to manoeuvre, and the balloon bursts.

At either end of a balloon, the polymers are relatively loose, which is why the colour is darker. Here the skewer can get through the gaps without bursting the balloon. The petroleum jelly assists the process by reducing friction.

Balloons are made of polymers

Polymers stretched

NO POP!

Jelly reduces friction

Stuff To Get
- A balloon
- A wooden skewer
- Petroleum jelly
  (e.g. Vaseline)

Let's Get Started

1. You're at a party, there are balloons everywhere, and the kids are out of control. There's only one way to silence them – skewer a balloon.

2. Nip into the kitchen and find a wooden skewer, make sure the tip is nice and sharp (be careful you don't poke your eye out with it). Rub some petroleum jelly on the tip and stroll nonchalantly back into the mayhem. Ask a small child if you can have their balloon. If they refuse, offer them a sweet or some money – that should work.

3. Shout 'Who wants to see some cool science' to attract everyone's attention or, failing that, cough loudly. Wait for silence and slowly push the pointed end of the skewer into the balloon. Make sure you start near the knot, where the balloon's colour is darkest.

4. As the audience holds its breath, push the skewer through the balloon and out at the opposite end, again where the colour of the balloon is darkest. Enjoy the relative quiet while everyone tries to work out how you did it.

Fire Extinguisher in a Jug

DANGER – ADULTS ONLY

The next time you want to play a neat trick on your friends why don't you try out this feat of blowing out a candle – without using your lungs! Not only will this make you look like a jolly clever fellow, but you'll also learn a magic bit of cool chemistry ...

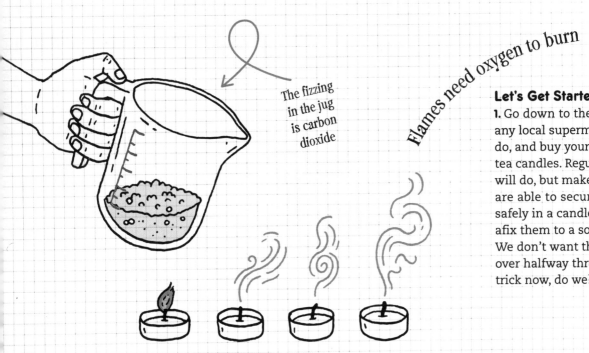

The fizzing in the jug is carbon dioxide

Flames need oxygen to burn

Let's Get Started

1. Go down to the shops, any local supermarket will do, and buy yourself some tea candles. Regular candles will do, but make sure you are able to secure them safely in a candle holder or afix them to a solid surface. We don't want them falling over halfway through the trick now, do we?

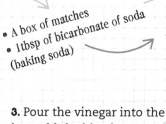

- Some candles (tea candles are perfect)
- A box of matches
- 1 tbsp of bicarbonate of soda (baking soda)
- 150ml (5¼ fl oz) white vinegar
- A large jug

2. In the kitchen, light the candles carefully and carry them into the room where your audience is waiting. Rather than blowing them out immediately, tell everyone you need to take a moment, and pop back into the kitchen.

3. Pour the vinegar into the jug. Add the bicarbonate of soda and watch the mixture bubble away.

4. Slowly and carefully carry the jug to the cake. Hold the jug over the candles and start to tip it, but not enough to allow the bubbling liquid to escape. Amazingly, the candles will go out.

The Science

Bicarbonate of soda has the chemical formula $NaHCO_3$, while vinegar is the common term for weak acetic acid (CH_3CO_2H) dissolved in water. Bicarbonate of soda reacts with acid (in this case the vinegar) to produce carbon dioxide gas (CO_2), which reduces the amount of oxygen in the air around the candles. As candles need oxygen to burn, the effect of reducing the amount of oxygen available will extinguish the flames. Carbon dioxide is also heavier than air and so tipping the jug above the candles will allow the carbon dioxide to replace the air around them and put out the flames.

Carbon dioxide fire extinguishers are the only type recommended for fires involving electrical equipment; they can also be used on flammable liquids.

Did You Know?

Dancing Matches

Moving objects with our minds, or without even touching them, is something we all wish we could do. Imagine summoning the TV remote control into your hand from the shelf where you left it, or a snack from the kitchen, all from the comfort of the sofa – you would be the ultimate couch potato. Sadly this trick does not involve telepathy like that, but it does use the science of sound waves to make matches dance.

Stuff To Get
- Two matches
- Two wine glasses
- Water

Let's Get Started

1. Fill each glass a quarter full of water. Make sure there is the same amount of water in each glass.

2. Balance the two matches on the rim of one of the glasses. Wet your finger and, on the glass with no matches, run your finger around the rim to make the glass ring.

3. Continue to make the glass ring and move it close to the glass with the matches. As you get closer, the matches will suddenly start to move.

Did You Know?

We hear because our eardrum is vibrated by the wave of sound coming from the source. Telephones use this same science, converting the sound wave into an electrical signal, transmitting it, and then converting it back into a sound wave.

Cheers!

Run your finger around the glass rim

Bring the glasses together – feel the good vibrations!

Watch as the matches dance together

The Science

Sound waves are powerful things. Everyone has heard about opera singers who can shatter glass with their voices and it is the same science at work here.

By running your finger around the rim of one glass, you are causing the bowl of the glass to vibrate, thereby creating a high-pitched sound. This sound is a wave and, because the other glass contains the same amount of water, and thus operates at the same pitch, it will pick up the wave and move in the same way. These vibrations cause the glass to move infinitesimally and it is these movements that make the matches dance.

Flying Eggs

PLEASE BE CAREFUL

Eggs are good for you, and any trick involving toilet rolls has to be worth the entrance fee alone. This amazing demonstration of the law of inertia is a real crowd pleaser.

Stuff To Get
- Three eggs
- Three cardboard tubes
- Three beaker-shaped glasses
- Water
- A tea tray (this must be totally flat on the bottom, with no protruding edge)

The Science
This trick is all about inertia, described by Sir Isaac Newton in his First Law of Motion. Because the cardboard tubes are relatively light they are pulled away with the tray while it is moving. As the eggs are heavier they require a greater force to move them in the same direction as the tray, and in this case they do not receive it, so instead they drop into the glasses below.

Did You Know?
Everything that is standing still has inertia, which means that it will not move unless forced to. Everything that moves has momentum, which means that it will not slow down or speed up or change direction unless forced to. There is no real difference between inertia and momentum, because everything in the Universe is moving. Things only appear to be still because they are not moving relative to something else.

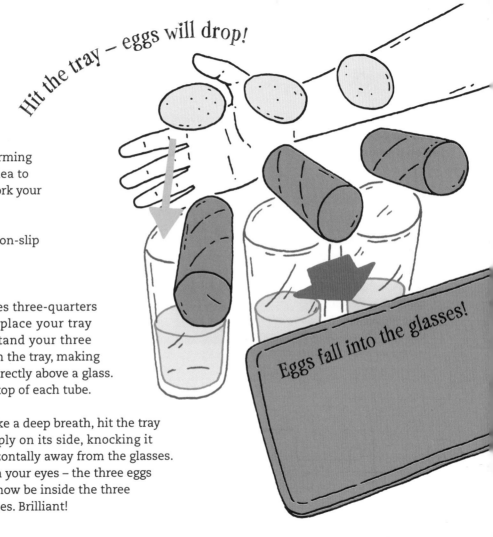

Hit the tray – eggs will drop!

Eggs fall into the glasses!

Let's Get Started

If you're nervous about performing this trick, it may be a good idea to just start with one egg and work your way up to the full omelette.

1. Arrange your glasses on a non-slip surface – in a triangle.

2. Fill all the glasses three-quarters full of water and place your tray on top of them. Stand your three cardboard tubes on the tray, making sure each one is directly above a glass. Put an egg on the top of each tube.

3. Take a deep breath, hit the tray sharply on its side, knocking it horizontally away from the glasses. Open your eyes – the three eggs will now be inside the three glasses. Brilliant!

Floating Arms

FOR EVERYONE

In dreams people often imagine they can fly and it can be quite depressing to wake up and realize that you can't. This trick really does give you the feeling that your arms are flying up, up and away. For a trick, this is a real doozy ... prepare for flight!

The Science

Any movement that takes place in your body is the result of an antagonistic relationship between opposite muscles.

In this example, when you are attempting to move your arms upwards, one muscle is tensing (the tricep), while another is relaxing (the bicep). When you let go of your jeans, because you have been pulling on them for so long, your muscles release restrained energy and seem to move by themselves.

Did You Know?

On average, muscles make up around 40 per cent of your body weight. There are over 630 of them. Muscles can't push, they can only pull, and as we have seen in this trick they often work in pairs so that they can pull in different or opposite directions.

Stuff To Get

• A pair of jeans

• Your arms

Let's Get Started

1. Stand with your legs shoulder width apart and let your arms hang down by your sides.

2. With your arms straight grab hold of a fistful of your jeans with each hand and start to pull outwards, away from your legs. Keep doing this as hard and for as long as you can.

3. When you think your arms are going to fall off because you've tried so hard, stop pulling, let go of your jeans, and relax your arms. Whoa – your arms will start floating upwards all by themselves!

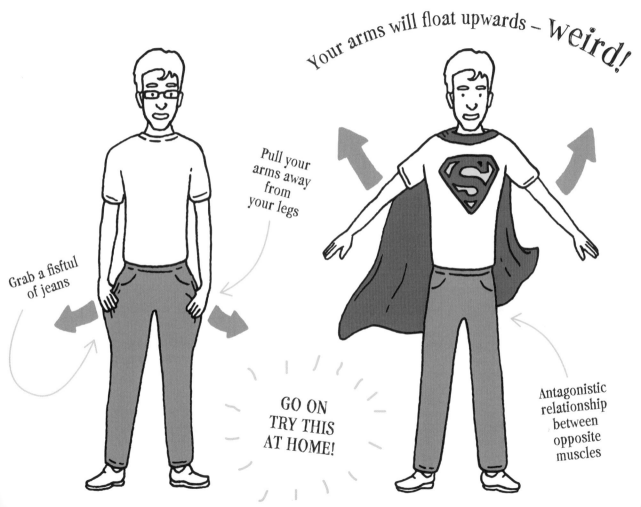

Your arms will float upwards – weird!

Pull your
arms away
from
your legs

Grab a fisftul
of jeans

GO ON
TRY THIS
AT HOME!

Antagonistic
relationship
between
opposite
muscles

41

Home-made Lava Lamp

PLEASE BE CAREFUL

They say that if you can remember the '60s you weren't there. Does that mean that if you can remember having a lava lamp in the '60s you didn't have one?

Stuff To Get

- A large bottle of cooking oil
- Water-based food colouring
- 500ml (17½ fl oz) water
- An effervescent pain-relief tablet (Alka-Seltzer or similar)
- A 2 litre clear plastic bottle

Let's Get Started

1. Put some Jimi Hendrix on the turntable, we're going back in time for a crazy night in.

2. Put 500ml (17½ fl oz) of water in the bottle and fill up the rest of the bottle with cooking oil. Wait for the oil and water to separate.

3. Add about ten drops of food colouring. Allow it to sink to the bottom and mix in with the water.

4. Drop in the tablets and stand back – you will suddenly forget why you ever bought a TV.

The Science

The science here is divided into three stages and they all need to work together to make the lava lamp work.

Stage 1: Water is denser than oil so it sinks to the bottom and the oil sits on top. Like the water, the food colouring is also denser than oil so it sinks through without mixing with it.

Stage 2: The effervescent tablets will only begin to fizz when they reach the water, at which point carbon dioxide is released into the water. This reduces the density of the water causing the coloured water to rise through the oil.

Stage 3: When it reaches the top the carbon dioxide is released from the bottle, the water sinks once again and we're back at Stage 2.

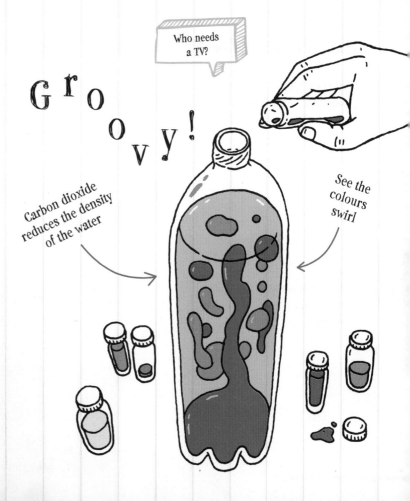

Groovy!

Who needs a TV?

Carbon dioxide reduces the density of the water

See the colours swirl

The Impossible Piece of Paper

PLEASE BE CAREFUL

Paper cuts can be both painful and annoying ... but to distract yourself, why not try this mind-boggling paper illusion?

Let's Get Started

1. Fold the piece of paper in half along its long axis and then open it out so it lies flat.

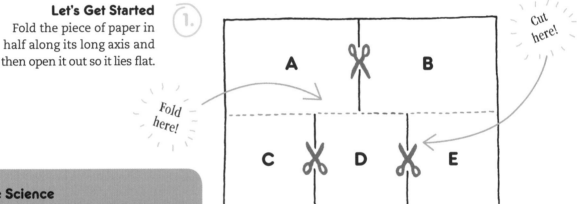

A   B

Fold here!

Cut here!

C   D   E

The Science

When you show off the final result, the viewer will be thinking of the piece of paper as a flat object, without recognizing the twist along its axis that makes this illusion possible. If the two sides of the paper were different colours it would be instantly solvable.

2.

Make a cut from the middle of the long edge on one side of the paper into the centre of the fold (dividing that side of the page in half). Along the long edge on the other side of the fold make two cuts from the edge to the middle of the paper (dividing that side of the page into three even segments).

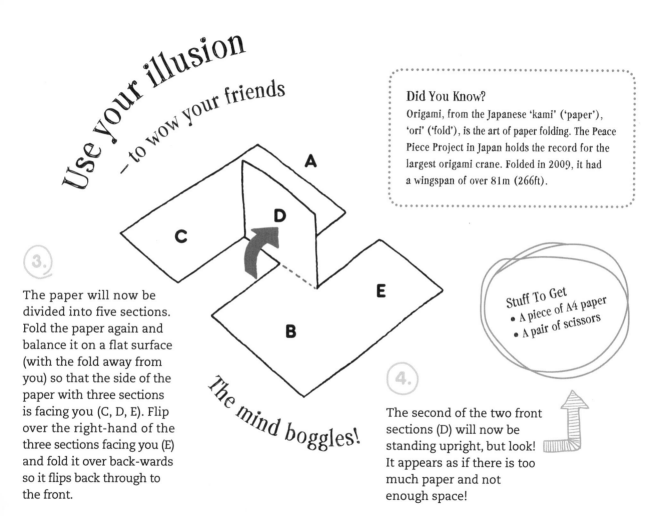

Use your illusion
– to wow your friends

A

D

C

E

B

The mind boggles!

Did You Know?
Origami, from the Japanese 'kami' ('paper'),
'ori' ('fold'), is the art of paper folding. The Peace
Piece Project in Japan holds the record for the
largest origami crane. Folded in 2009, it had
a wingspan of over 81m (266ft).

Stuff To Get
• A piece of A4 paper
• A pair of scissors

3. The paper will now be
divided into five sections.
Fold the paper again and
balance it on a flat surface
(with the fold away from
you) so that the side of the
paper with three sections
is facing you (C, D, E). Flip
over the right-hand of the
three sections facing you (E)
and fold it over back-wards
so it flips back through to
the front.

4. The second of the two front
sections (D) will now be
standing upright, but look!
It appears as if there is too
much paper and not
enough space!

45

Levitating Water

FOR EVERYONE

Every now and then you see something that seems to defy science. A bee should not be able to fly but it does and this little piece of magic is just as surprising. Is it really possible to turn a glass of water upside down without the water coming out? Yes it is – and here's how.

Stuff To Get

- A glass
- Water
- A piece of stiff card (e.g. a postcard)

Let's Get Started

1. Fill the glass up to the brim with water.

2. Place the card over the top and tap it in the middle to make sure it has formed a seal. Hold the card with one hand and turn the glass upside down.

3. Making sure the glass is totally straight, in other words the card is parallel to the ground, let go of the card, and then … nothing happens! The card holds, the water stays in the glass.

Did You Know?

The higher you are, the less air pressure there is. This is unhealthy for humans because they are deprived of oxygen, so in an aeroplane for example, passengers are provided with artificial air pressure. If you are at 31km (19 miles) above the surface of the Earth, the pressure is $1/100$th of that at sea level. And at 100km (62 miles), you are already in outer space, and the air pressure falls to zero …

Simple but amazing!

Let go!

Atmospheric pressure at work

Make sure a seal has formed – give the glass a tap

The card stays stuck in place!

The Science

It is air pressure that is at work here. The atmosphere exerts about 1kg per sq cm (14.7lb per sq in) at sea level. Because it is a gas, air not only pushes downwards but also upwards from the bottom and from the sides. The card stays in place because the pressure of the air molecules pushing up on the card is greater than the pressure of the water pushing down.

Living on the Edge

PLEASE BE CAREFUL

An audience is excited by danger – the fear that someone might get hurt. When we watch a man walking on a tightrope we don't want him to fall, but it's the thought that he might which keeps us watching. This trick allows you to create that tension using just a couple of forks and a toothpick.

Let's Get Started

1. Interlock the tines (or prongs) of the forks so they are stuck together.

2. Balance the central point between the two forks along their edge on your finger to locate the balancing point. Once you have found it push the toothpick a little way through a gap in between the tines.

3. Then balance the toothpick on the edge of the glass (you might need to slide the toothpick back and forth until you find the exact balancing point), let go, and amaze your audience!

... et voila!

Stuff To Get
- Two forks
- A toothpick
- A glass

Did You Know?
It was the English Mathematician Sir Isaac Newton who, in 1687, coined the term 'gravity' from 'gravitas', the Latin word for heaviness.

Balance the toothpick on rim of the glass!

The Science
The science in this trick is based on the concepts of centre of gravity and stability. The centre of gravity of an object is the point at which an object's weight is evenly distributed. When you are balancing the fork, the centre of gravity is directly below the point where the toothpick rests on the rim of the glass (the pivot point).

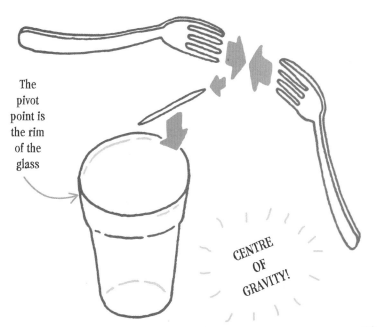

The pivot point is the rim of the glass

CENTRE OF GRAVITY!

Hot Ice

PLEASE BE CAREFUL

You definitely wouldn't want to use this ice to cool your gin and tonic because it is actually hot and of course it's not really ice. But the effect is really cool.

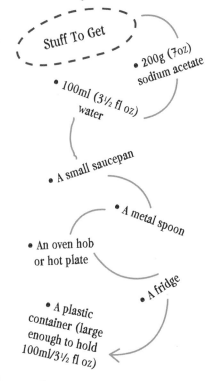

Stuff To Get

- 200g (7oz) sodium acetate
- 100ml (3½ fl oz) water
- A small saucepan
- A metal spoon
- An oven hob or hot plate
- A fridge
- A plastic container (large enough to hold 100ml/3½ fl oz)

Let's Get Started

1. Place the water in a saucepan and bring almost to the boil. Start adding the sodium acetate to the water, without letting it boil, and stir constantly with a metal spoon. Keep adding the sodium acetate until no more will dissolve – you will know you have reached this point when, on adding more to the solution, it just sinks to the bottom.

2. Pour the solution, but none of the undissolved crystals, into a plastic container, and place in the fridge for an hour.

Wait patiently by the fridge. Or do your homework. Or have a sandwich. Whatever you can think of that will take an hour!

HOT ICE!

Nucleation site – HERE!

Crystallization is happening

3. After one hour, take the container out of the fridge, place it on a flat surface, gather your friends around and blow really hard on your finger. Then, as a hush descends, touch the solution.

4. When you touch it, the solution will instantly crystallize, apparently turning it into ice, but, if you touch the container, it will feel warm. Hot ice, anyone? Well, warm ice – but still extremely weird! Once the amazement wears off, discard the hot ice safely in the bin.

The Science

Sodium acetate, also known as sodium ethanoate (CH_3COONa), is the sodium salt of acetic acid (household vinegar is a weak solution of acetic acid). Dissolving sodium acetate in near-boiling water creates a supersaturated solution. This solution can cool without forming crystals. When you touch the liquid you create a nucleation site, and it is one of the properties of molecules in a supersaturated solution that they will aggregate and crystallize at the nucleation site, and in this case they form hot ice right before your eyes!

Did You Know?

This experiment has practical applications in the real world and this is exactly the same science that is used in hand-warmers – the creation of supersaturated solutions without forming crystals to produce warm, dry air. Sodium acetate is also the primary flavouring in salt and vinegar crisps.

Egg in a Bottle

DANGER – ADULTS ONLY

A ship in a bottle takes a lot of skill, time and attention to detail. This egg in a bottle trick is slightly easier, although the relative scarcity of the glass milk bottle (and indeed milk men) has made this stunning piece of magic a little harder, although once you find one it is a breeze.

The Science

Atmospheric pressure exists all around us. Initially the pressure inside the milk bottle is the same as outside. When the paper begins to burn, the heat causes the air pressure in the bottle to increase and some of the air escapes past the egg. Once the paper stops burning, the air inside the bottle begins to cool. Ordinarily air would come rushing back into the bottle to equalize the air pressure, but, as the egg is in the way, the pressure builds up until it becomes so great that the egg is sucked into the bottle.

Stuff To Get

- A glass bottle with a wide mouth, e.g. a milk bottle
- A hard-boiled egg
- A strip of paper
- A box of matches

Pressure builds up inside the bottle!

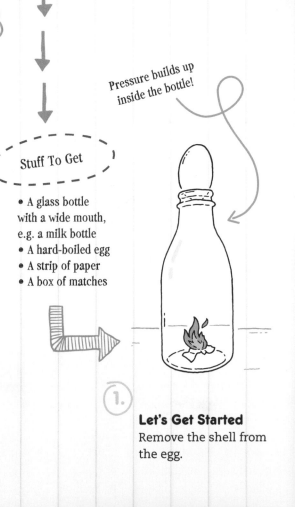

1.

Let's Get Started

Remove the shell from the egg.

2.

Light the paper using a match and drop it, while it is burning, into the bottle. Put the egg on top of the bottle.

 *Wait for it!*

Good luck getting the egg out!

3.

Once the paper burns out, wait a few moments, and watch as the egg is sucked into the bottle.

Watch the paper burn!

Did You Know?

The Earth's atmosphere is pressing against you with a force of 1kg per sq cm (14.7lb per sq in). The force on 1,000 sq cm (a little larger than 1 sq ft) is about one tonne! But remember the air inside our bodies balances out the pressure outside so we do not resemble that crushed can we talked about earlier.

The Airborne Ping-pong Ball

We would all love to be able to fly, but it's just not possible. Some of us would even like to be able to float in mid-air above a hair-drier, but that's not possible either. However, we can do that to a ping-pong ball.

The Science

There are two interesting things happening here. The first is gravity: the air produced by the drier pushes the ball up while gravity is doing its best to pull it down. The ball finds the point at which these two forces balance each other out and stays there – levitating. The second is air pressure: fast-moving air creates low air pressure and the stream of air coming out of the hair-drier creates a column of low pressure around the ball. The higher air pressure outside this column prevents the ball moving out of the stream.

It is the science of air pressure that allows aeroplanes to fly. The wing of an aeroplane is curved on top. When air hits the front of the wing, the air that goes over the top of the wing travels faster than the air that goes underneath it. This means that the air pressure above the wing is lower than below the wing, creating an uplift – we have lift-off!

Did You Know?

Let's Get Started

1. First check that no one is getting ready to go out – you may get shouted at. If the coast is clear, plug in the hair-drier and turn the heat setting down to cool.

2. Make sure the nozzle of the hair-drier is pointing straight up and place the ball on the nozzle.

3. Turn the hair-drier on and watch the ball as it floats in mid-air.

Hot air pushing ball up!

Gravity pulling ball down

Ball floats in the air like

MAGIC!

Nine Dots, One Line

FOR EVERYONE

This is a classic brain-teaser and is guaranteed to turn all your friends against you when you show them how easy it is to solve.

Think outside the box!

- A piece of paper
- A pencil
- A book

Stuff To Get

The Science

There is something in our genetic make-up, or our basic human instinct, that means we want to keep things simple and tidy, and as a consequence most people who attempt this trick will not venture outside the range of the dots. It just goes to show that a good scientist has to think outside the box.

Get your friends to try too – it'll drive them dotty!

Brain tease!

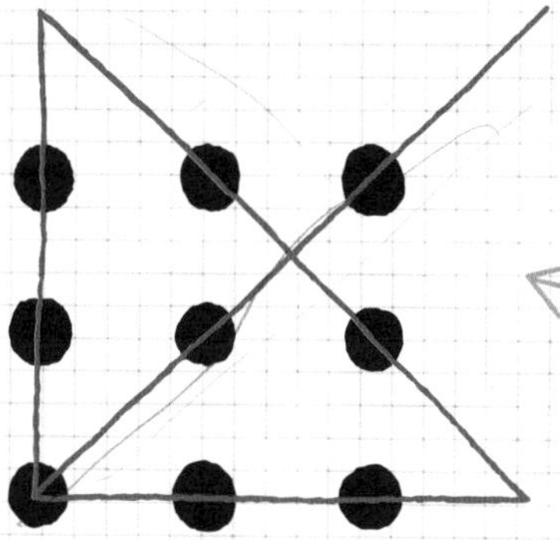

Did You Know?
If the dots are big enough, it is possible to solve this in three lines. Can you see how?

Illustration Answer on page 111

Let's Get Started

1. Draw nine dots on a piece of paper as shown above. Then ask your friends to draw four lines that go over each dot, without taking the pen off the page and without going over any dot twice.

2. Leave the room for a while and read your book, while your friends try to solve the puzzle. When you start to hear screams of frustration go back in and show them what to do. You'll find the answer on page 111.

The Breath Test

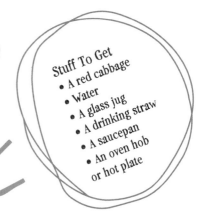

PLEASE BE CAREFUL

Stuff To Get
- A red cabbage
- Water
- A glass jug
- A drinking straw
- A saucepan
- An oven hob or hot plate

Nature provides the raw ingredients for all science. Everything is there if you know where to look. Red cabbage, a tasty treat, can also be used in the lab to indicate the pH content of other substances.

Let's Get Started

1. Boil a few leaves of the red cabbage in a pan of water. Pretty soon the water will turn pink.

2. Strain the cabbage and pour the (now pink) water into a glass jug.

3. Using the straw, blow into the cabbage juice and watch it brilliantly change colour: just the power of your breath will make it turn redder and redder!

The Science

Red cabbage contains a water-soluble pigment molecule called flavin, which is an acid/base (alkali) indicator. This means that depending on what colour it changes to you can tell the pH value of a substance mixed with it: acidic solutions will turn red; neutral solutions will result in a purple colour; base (alkali) solutions will appear greenish-yellow. As the carbon dioxide in your breath is acidic, when you blow into the cabbage water, it shows as a red colour.

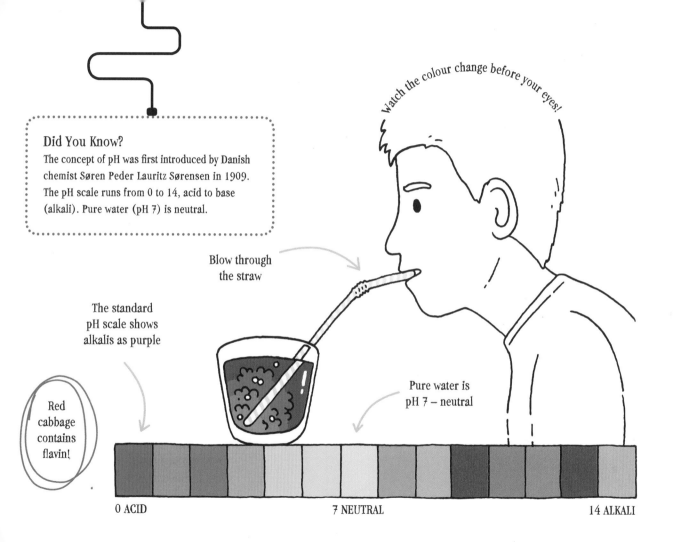

The Magic Straw

FOR EVERYONE

Straws are cool anyway, they allow you to drink liquids without having to pick up the bottle, and this amusing trick appears to bring the straw to life as you can make it spin without touching it.

Let's Get Started

1. Balance the straw on the lid of a bottle and ask one of your audience to move it without touching or blowing it – of course they won't be able to.

2. Pick up the straw, wrap a piece of cloth around it (your T-shirt will do), and with a sharp motion pull the straw through it. Then place it back on the bottle.

3. Now put your hands near the straw and, as if by magic, it will start to move away from your hands.

- A plastic drinking straw
- A bottle

Stuff To Get

Did You Know?
When you rub the head of someone who has long hair with a balloon, the hairs become positively charged. If the hair is light enough (blond hair is best as it is finer), the individual hairs will repel one another and stand upright, until the charge comes back into balance.

The Science

Static electricity is the invisible secret behind the movement of the straw. There are two stages of static electricity at play here. The first is when you pull the straw through your T-shirt: this causes the electrons in the straw to move to your T-shirt producing an imbalance in the electric charge on the surface of the straw. The second stage occurs when you put your hands close to the straw: the difference in charge between your hands and the straw causes the straw to move away.

Use your superpowers to move the straw around – without touching it!

Jedi mind trick!

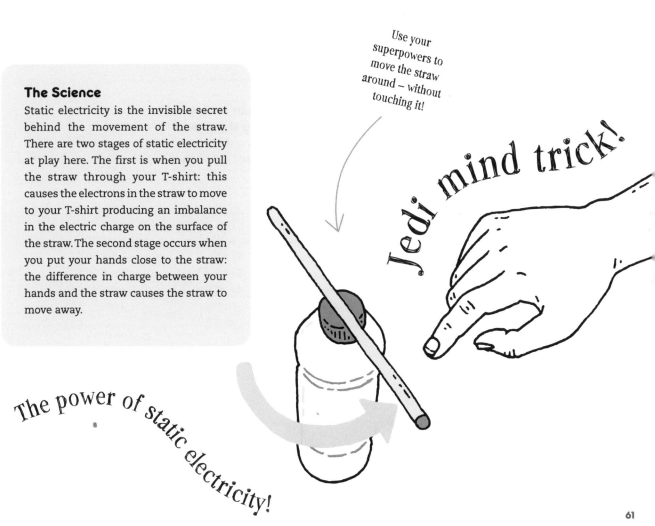

The power of static electricity!

Let There Be Fire

You're stuck in the jungle, you're tired, and you just want to get a nice fire going before the sun goes down. Unfortunately, your matches got wet in the last stream you waded through. Luckily, you have a can of drink to quench your thirst, but once it is finished you should really clean your teeth to get rid of that sugar. Hold on a minute – you now have the ingredients for a fire.

Let's Get Started

1. Put some toothpaste on the piece of cloth or paper and rub the base of the can with the toothpaste. Keep rubbing until the whole base of the can is really shiny and you are able to see your reflection in the surface. Then gather some kindling.

2. Point the bottom of the can towards the sun to make a small, focused ray of light, and then aim this at the kindling.

3. Hold it still and after a few minutes (depending on the time of day and the strength of the sun) you will see some curls of smoke, followed by ignition and fire.

Did You Know?
The term photon was coined by American scientist Gilbert Lewis in 1926, but the modern concept of the photon, namely that it has the dual characteristics of waves and particles, and thus can carry energy, was developed gradually by Albert Einstein.

Stuff To Get
- A soft-drink can
- Whitening toothpaste (this is the most abrasive)
- A piece of cloth or paper
- Kindling
- Sunlight

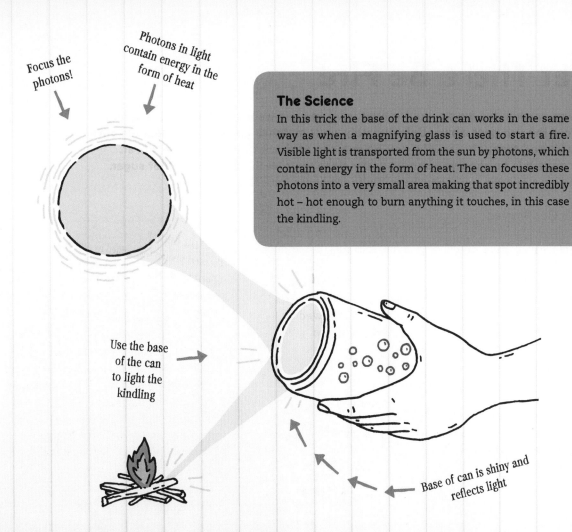

Focus the photons!

Photons in light contain energy in the form of heat

The Science

In this trick the base of the drink can works in the same way as when a magnifying glass is used to start a fire. Visible light is transported from the sun by photons, which contain energy in the form of heat. The can focuses these photons into a very small area making that spot incredibly hot – hot enough to burn anything it touches, in this case the kindling.

Use the base of the can to light the kindling

Base of can is shiny and reflects light

The Hidden Rainbow

FOR EVERYONE

Black is the coolest of colours but in fact it is not a colour at all: it is made up of all the other colours in the light spectrum and this fun trick shows that to beautiful effect.

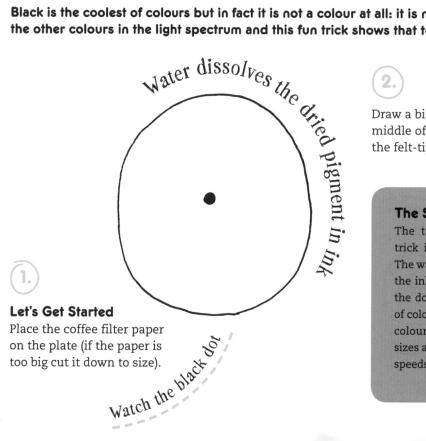

Water dissolves the dried pigment in ink

Watch the black dot

1.

Let's Get Started
Place the coffee filter paper on the plate (if the paper is too big cut it down to size).

2.

Draw a big black dot in the middle of the paper using the felt-tip pen.

The Science
The technique demonstrated in this trick is called paper chromatography. The water dissolves the dried pigment in the ink and as it travels outwards from the dot the ink is separated into bands of colour. Because the molecules of each colour in the spectrum are different sizes and weights they travel at different speeds and over further distances.

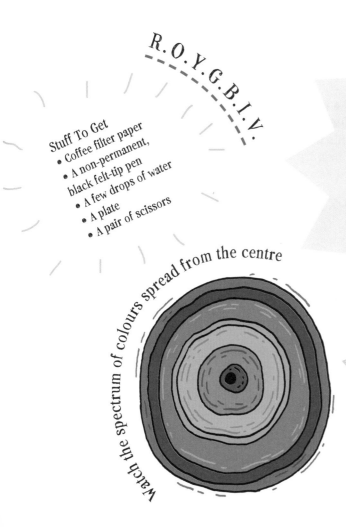

R.O.Y.G.B.I.V.

Stuff To Get
- Coffee filter paper
- A non-permanent, black felt-tip pen
- A few drops of water
- A plate
- A pair of scissors

Watch the spectrum of colours spread from the centre

Chromatography, from the Greek 'chroma' ('colour'), 'graphein' ('to write'), was invented by Russian scientist Mikhail Tsvet in 1900. It was first used for the separation of plant pigments, but new types have since been developed and it is now possible to perform chromatography on gases and liquids.

Did You Know?

THE LIGHT SPECTRUM
Red, Orange, Yellow, Green, Blue, Indigo, Violet

3.

Drip a few drops of water on to the dot, stand back, and watch as a rainbow spreads out from the dot.

The Cola Can-can

FOR EVERYONE

For some, this sugary fizzy drink can only truly be enjoyed from a glass bottle but for this particular amazing trick it has to be a can – what the bottle can't do, the can can. You'd think with 1.4 billion cans sold every day that everyone would know this trick, but amazingly they don't. So, recycle wisely and, next time you finish a can of your favourite fizz, try out this nifty bit of physics on your friends.

Let's Get Started

1. Pick up your drink can and open it. Quench your thirst by gulping it down in one – be careful not to let it go up your nose. You could sip it gently, but where's the fun in that?

2. When the can is empty, carefully pour 100ml (3½ fl oz) of water into the can and balance it on its rim on a flat surface. (Make sure the hole is facing away from the direction you are tipping the can to avoid any spillages if the can does fall over.) Watch in amazement as it balances all by itself at a 45° angle. Magic! Wave your hands around the can to show that there are no strings attached.

3. As if that is not enough, next with a gentle nudge of your finger, watch the can spin round and round, right round baby like a record baby, right round.

The Science

Trial and error led to the discovery that 100ml (3½ fl oz) was the perfect amount of water to get the can to balance on its edge at a 45° angle. A little too much and it will topple over, a little too little, it falls back on to its base. It's a simple bit of physics, but it's beautiful.

Stuff To Get

• A finger that can nudge

• 100ml (3½ fl oz) water

• A standard fizzy drink can, 330ml size with a rim on the base

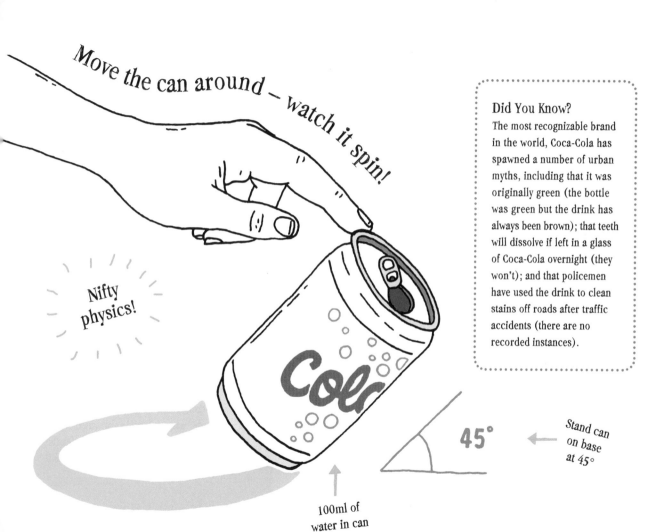

Move the can around – watch it spin!

Nifty physics!

Cola

45°

100ml of water in can

Stand can on base at 45°

The Disappearing Coin

This is not a trick that tries to explain where your money goes when the stock exchange collapses and the bankers buy bigger cars. That is indeed beyond science. It does, however, demonstrate some simple science and some amazing magic.

Let's Get Started

1. Ask a friend, an acquaintance, or maybe just someone you meet in the street, to lend you a coin.

2. Place the coin under a glass and ask your friend/acquaintance/mark if they can still see the coin. (If they say no at this point you are in trouble. Hopefully they'll say yes.) Fill the glass up with water and place the saucer on top.

3. Ask them again if they can see the coin. They won't be able to – it has disappeared!

Stuff To Get
- A coin
- A glass
- A saucer
- Water

Did You Know?
Eyes are amazing. Your retina contains 120 million rods for 'night vision', and 8 million cones that are colour sensitive and work best under daylight conditions. Under the right conditions, the human eye can see the light of a candle at a distance of 22.5km (14 miles).

Fool your friends with light refraction!

Pop the coin under the glass!

Your eyes won't believe it!

The Science

Our eyes do not actually 'see' things by themselves: the light receptors in our eyes allow us to see light coming off the surface of objects thereby revealing what it is we are looking at.

When the coin is under the empty glass, the light reaches us quite easily and so the coin is still visible. With water in the glass and the saucer on top, the light refracts and its direction is altered so that it cannot travel in a straight line to our eyes. This makes it appear as if the coin has disappeared. If you take the saucer off the top of the glass, you will see the coin in the surface of the water.

The Ultra-strong Straw

Friends will often argue about whose turn it is to buy the next round of drinks or pay the bill for dinner. This trick will ensure your friends pick up the tab as you show them how to pick up a bottle using just a straw.

Did You Know?

The first manufactured drinking straws were made of paper. In 1888, American inventor Marvin Stone patented a spiral winding process to produce the straws. His prototype was made using strips of paper that were wound around a pencil and then glued together. He then worked with coated paper so that the straws did not become wet while they were being used. Almost 50 years later another American, Joseph B. Friedman, invented the flexible drinking straw.

The power of mechanics

PLEASE BE CAREFUL

Straw acts as a lever

Lower the straw into the bottle — bendy bit first!

Fulcrum!

Stuff To Get

• A drinking straw

• A bottle

Let's Get Started

1. When you order your next drink, ask for a bottle of beer and a straw. Then make a bet with your friends that you can pick up the bottle with the straw.

2. Bend the straw about 5cm (2in) from one end, and holding the other end push the straw into the bottle with the bend entering first.

3. Slowly lower the straw into the neck of the bottle until the short end unfolds and catches in the neck. The short end should now be wedged in the neck.

4. Lift the bottle, smile your best smile and say, 'Your round, I believe?'

The Science

This trick illustrates some basic mechanics. The bent straw inside the bottle acts as a lever, which is used to pick up the bottle. The bend in the straw is the fixed support or fulcrum, which is used to transmit effort and motion; the short section of the straw between the fulcrum and the inside top of the bottle is the load arm; and the long section of the straw, between the fulcrum and the lifting force (in other words, you), is the effort arm.

Multi-coloured Milk

It's really exciting when you see a rainbow in the sky but, sadly, this just doesn't happen often enough. This bit of chemistry allows you to create a rainbow in a bowl in your kitchen with materials you can find around the house.

Stuff To Get

- Food colouring – the more colours you have the better
- Full-fat milk
- Washing-up liquid
- A shallow bowl – a large, shallow bowl with a wide base will work the best

Did You Know?

It takes over 70 squirts from a cow's teat to collect just 1 litre (1¾ pints) of milk.

Let's Get Started

1. Pour some milk into the bowl, enough to cover the base of the bowl, and about 1cm (²⁄₅ in) deep.

2. Then drip about four drops of each different colour of food colouring, evenly spaced around the edge of the bowl.

3. Gently drip a few drops of the washing-up liquid into the centre of the bowl, then stand back and watch in amazement as the colours rush across the surface of the milk and start to mix together in a fantastic, swirling display.

The Science

Detergents such as washing-up liquid like to join together water and fat (or grease), which is why they are good at cleaning dishes. Here the washing-up liquid tries to get the water and fat in the milk to mix together and as it does so it creates swirls and eddies in the milk that cause the different colours to mix together.

The Paper Cut

We've all done it: you're at home or in the office and you reach for a piece of paper and the edge slides against your finger – owwwwwch! Paper cuts are the most painful cuts of all, no one will argue with that. But did you know you can magically cut a pencil in half with a piece of paper? Well, read on ...

The Science

The science here is about distraction and the way our brains works. Our brains don't always process everything that flashes before us, and in this trick your 'victim' will be watching the pencil, and their fingers. As everything happens so fast, they won't see that your finger is helping you to break the pencil – as long as you get the timing right of course.

Did You Know?

Researchers have calculated that each of our eyes transfers information to our brains at about the same speed as a standard broadband Internet connection. Impressive, huh? But according to scientists our neurons could work a lot faster than this. The reason they don't is that our brain makes up 2 per cent of our body weight but needs 20 per cent of our energy to keep it working. The ganglion cells in our retina are divided into 'brisk' (that bring you the most important information) and 'sluggish' (that transmit less important data), and in order to conserve energy the sluggish cells carry out the majority of the work.

Let's Get Started

1. Get a friend to lend you some paper money and then ask, 'If I can cut a pencil in half with this note, can I keep it?' When they say yes, get to work – don't give them a chance to change their mind.

2. Carefully fold the note in half lengthways. With a flourish, offer your friend the pencil to hold. They need to hold each end firmly.

3. Hold the note between your thumb and index finger at one end. Raise the note above your head and quickly whip it down towards the pencil. As you reach the pencil, flick out your index finger and use it to help you break the pencil.

4. When the note and your finger have split the pencil, bend your finger back. Your friend will be left with half a pencil in each hand, while you will be left with the money and a big smile.

Stuff To Get
- Some paper money
- A pencil

Use your index finger to break the note!

Money to Burn

DANGER –
ADULTS ONLY

Money doesn't grow on trees, and it can't buy you love – so maybe money doesn't have that much going for it. But just try borrowing some paper money from a friend for this trick and see how they react. Or how about performing it for your kids? It should teach them about the value of money.

In the USA, it is illegal to destroy bank notes. So be careful! If this trick goes wrong you could be fined and/or imprisoned for up to six months.

Did You Know?

The Science
Scientifically, this trick is actually quite simple to explain. The alcohol present in the surgical spirit is what is burning here, not the paper. But, and this is the important bit for the fate of the money, the heat of the flames is not great enough to evaporate the water. The water protects the note from the flames and, when all the alcohol has been burned, the water puts out the fire.

Some paper money – try borrowing someone else's in case the trick goes wrong!

Water protects the money from the flames

The money isn't burning – the alcohol in the surgical spirit is!

Surgical spirit

Kitchen or barbecue tongs

A tumbler

A lighter

Water

Let's Get Started

1. Ask someone if you can borrow some paper money. When they have finally agreed soak it in a tumbler in which you have a mixture of half water and half surgical spirit. Announce it is now flammable and it would be dangerous for your friend to try to retrieve it.

2. Take the note out of the tumbler using the tongs and hold it at arm's length. (It is a good idea to do this trick outside, on a day with little or no wind.) Set fire to the note with the lighter, watch it burn brightly.

3. Watch your friend look on in horror until … the flame goes out leaving the note intact, if a little damp.

Be careful with your fingers

LIGHT THE MONEY!

The Flying Man

FOR EVERYONE

Is it a bird? Is it a plane? No, it's just someone at home who has developed the ability to fly thanks to this simple but impressive (to the kids at any rate) optical illusion.

Did You Know?

'Trompe-l'oeil', from the French for 'deceive the eye', is an art technique that tricks the observer into thinking that a painted object or scene is actually real. Also known as illusion paintings, the technique has been used throughout history, often in grand household interiors on walls and ceilings, and in theatrical set design.

Let's Get Started

1. Stand side-on to the mirror, with half your body away from your audience, and half your body reflected in the mirror.

2. Lift up your visible leg and arm – from the audience's point of view, it will look like you have lifted both legs off the ground. Remember not to actually lift both your legs, as you will fall over!

Stuff To Get
- A long mirror (it must be at least as tall as you)
- Yourself

Trick your brain!

The Science

The brain is very clever and attempts to make sense of the images that are sent to it. This is why, when we watch a film at the cinema, we don't see a series of still images, we see real movement. Sometimes though, the brain is too clever for its own good. In this trick, at first glance the image is processed as a whole person, rather than what it really is, half a person reflected in a mirror.

The reflection fools your eyes

Stretch out your arms, lift up your leg!

The Tablecloth Trick

This is a classic party trick – whipping a tablecloth off a table and leaving the crockery, cutlery and glasses standing where they were. When it works it's amazing, when it doesn't it's smashing, literally. You've seen it done – now try it out for yourself!

The Science

Our old friend inertia is at work here. Sir Isaac Newton described inertia as the tendency for an object at rest to remain at rest until a force acts upon it. In this case, the items on the table will not move unless moved by another force. When you pull the tablecloth, the force that acts on the items is friction, but as the tablecloth is slippery these forces are small, allowing the cloth to be pulled out from under the objects without dislodging them.

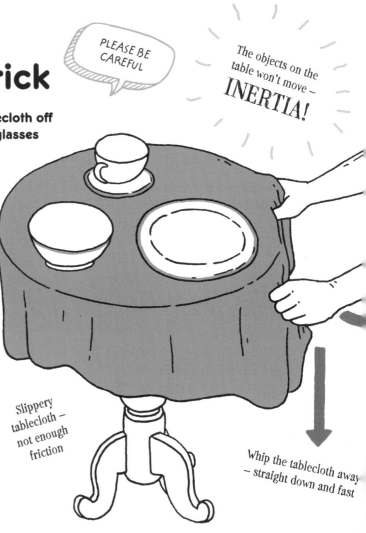

PLEASE BE CAREFUL

The objects on the table won't move –
INERTIA!

Slippery tablecloth – not enough friction

Whip the tablecloth away – straight down and fast

Let's Get Started

1. You've just finished dinner and you want to wash the tablecloth without clearing the table. Move everyone away from the table – just in case.

2. Hold the edges of the tablecloth really tightly with two hands.

3. Whip the cloth away – the knack is to pull it straight down as fast as possible.

NB It is a good idea to practise this trick on a partially laid table to begin with to build up your confidence.

Did You Know?
British juggler and comedian Mat Ricardo has taken this trick one step further: as well as pulling the cloth off the table leaving everything intact, he can also put it back on again without disturbing the objects. Now that's magic (and science)!

Ta-da!

The Flame Jumper

In 'Star Trek' the transporter allowed the crew of the USS Enterprise to jump across space. Science itself has not yet caught up with science fiction but it can make fire jump, and you can try it too. Sounds impossible but ...

DANGER – ADULTS ONLY

Stuff To Get
- A candle
- A match

Did You Know?

The friction match was invented in 1827 by an Englishman, the apothecary John Walker. The phosphorus match was invented in France in 1831 by French student Charles Sauria. Today, approximately 500 billion matches are used each year and about 200 billion of these come from matchbooks.

Let's Get Started

1. Strike the match and use it to light the candle. Keep the match lit.

2. Blow out the candle. Move the lit match into the stream of smoke coming from the extinguished wick roughly an inch away from the top of the candle.

3. The flame will 'jump' down the smoke and relight the candle.

KEEP THE MATCH LIT!

The candle will relight!

The flame will jump!

The Science

Candles are made from paraffin wax, and when they burn it is the vapour that is burning. In this trick, when you blow out the candle the smoke is also paraffin vapour, and so when you put a lit flame into that vapour it will burn all the way back to the source, the wick, and relight it.

Candles are made of paraffin wax

The Unbreakable Egg

PLEASE BE CAREFUL

You can't make an omelette without breaking eggs but what if you can't break the egg in the first place? Challenge a friend to break an egg and they will look at you as if your brains are scrambled. While they are laughing, take some money out of your pocket, put it on the table, challenge them to a bet, and show them this trick.

A normal egg

Now, wrap it in clingfilm!

Stuff to Get
- An egg
- Cling film – although if you are feeling extra confident you can try the trick without it

Let's Get Started

1. Wrap the egg in the cling film and hand it to your friend. Ask them to place it in their palm and close their hand so that their fingers are completely wrapped around the egg.

2. Then ask them to squeeze the egg by applying even pressure all around the shell. Try as they might, they will not be able to do it, so pick up your money (and theirs) and leave.

The Science

While the shell of an egg is fragile, the shape of an egg is one of the strongest architectural forms. When you squeeze the egg in your hand, pressure is applied evenly all over the shell, rather than focusing it at a single point, and it will not break. However, if you do apply pressure at a single point, as you do when you crack an egg on the side of a bowl, the shell breaks easily.

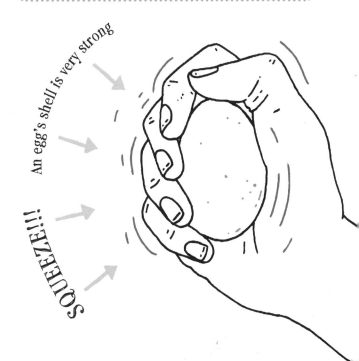

An egg's shell is very strong

SQUEEZE!!!

The Visible Sound Wave

PLEASE BE CAREFUL

We would all like to hear what someone else is thinking, but while that would be great it is, sadly, the stuff of science fiction. But this fun piece of physics allows us actually to 'see' noise.

The Science

Sound waves exist as variations of pressure in the molecules that surround us in a medium such as air. When someone speaks it causes vibrations in the molecules in the air, which causes your eardrum to vibrate, and the brain then interprets this as sound. Because molecules are so small we can't see this happening but, with this trick, the sound waves disturb the cornflour, which we can see.

Stuff To Get
- Cornflour
- Water
- Bin liners or cling film
- A sound system
- Some great tunes

Let's Get Started

1. Take the cover off one of your speakers, lay it down on its back and carefully wrap it in cling film or a bin liner (or two). Remember, your insurance is unlikely to cover you if this goes wrong!

2. Mix some cornflour and water together to create a runny paste (it is interesting to experiment with different thicknesses).

3. Put on a banging tune, turn it up loud, and pour the cornflour paste onto the covered speaker: the sound waves will disturb the mixture and make amazing shapes.

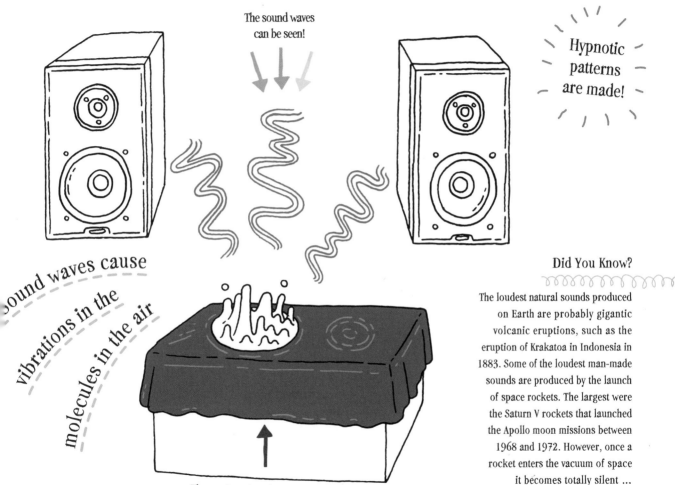

The sound waves can be seen!

Hypnotic patterns are made!

Sound waves cause vibrations in the molecules in the air

The music makes the mixture move!

Did You Know?

The loudest natural sounds produced on Earth are probably gigantic volcanic eruptions, such as the eruption of Krakatoa in Indonesia in 1883. Some of the loudest man-made sounds are produced by the launch of space rockets. The largest were the Saturn V rockets that launched the Apollo moon missions between 1968 and 1972. However, once a rocket enters the vacuum of space it becomes totally silent …

Witches' Brew

PLEASE BE CAREFUL

This trick is definitely one way to impress the kids: nothing is going to make you cooler than turning some everyday kitchen ingredients into a fizzing, expanding mass.

Stuff To Get
- A glass
- 2 tsp bicarbonate of soda (baking soda)
- White vinegar
- A few drops of food colouring
- Water

Let's Get Started

1. Fill the glass with water until it is about one quarter full. Add a couple of teaspoons of bicarbonate of soda, followed by a few drops of food colouring (red is best, it looks more scary).

2. Look at the potion, glance quizzically at your audience, and say, 'Why is it not working? Oh yes, the final ingredient ...'

3. Start to pour in the vinegar while issuing forth an evil laugh – the potion will start to bubble and expand.

Did You Know?

The first fizzy drinks were created by adding bicarbonate of soda to lemonade (the juice of lemons mixed with water and sugar). The citric acid in the lemon juice would react with the bicarbonate of soda to produce the 'fizz'.

Today, fizzy drinks are made by passing pressurized carbon dioxide through water. More carbon dioxide dissolves in the water at high pressure than at standard atmospheric pressure. When the bottle is opened, the pressure is reduced, and so dissolved carbon dioxide rushes out of the bottle, which causes the formation of the bubbles.

Neutralization reaction occurs!

Fizzing!

Bubbling!

A perfect Halloween concoction!

The Science

Why do the vinegar and bicarbonate of soda make such a great combination? Vinegar is a weak solution (about 5 per cent) of acetic acid (CH_3COOH). Bicarbonate of soda or sodium bicarbonate ($NaHCO_3$) is a base or alkali. When the two are mixed, a neutralization reaction occurs, which releases carbon dioxide gas (CO_2). In addition, the carbon dioxide that is created takes up more space than the bicarbonate of soda and liquid vinegar, hence the bubbling, fizzing, expanding potion.

The Liquid Stack

PLEASE BE CAREFUL

If someone asked you to create a pile of liquids you might suspect that they have drunk too much of a certain liquid themselves, but think again. Liquids with different densities do not mix and so you can in fact pile them high, and this trick utilizes that fact to stunning and beautiful effect.

Let's Get Started

1. Pour the liquids into the container in the order shown left, starting with the honey. Slowly and carefully add the next one, and so on. It is better to tip the container when pouring so the liquid runs down the side of the container.

2. Repeat this process until all the liquids have been poured into the container.

3. When you have finished you will see that all the liquids are sitting on top of one another while remaining completely separate from those above and below.

Did You Know?

The B-52 is a classic cocktail that makes beautiful use of layered liquids. At the bottom is a coffee liqueur, followed by Irish cream, and topped with a layer of orange liqueur.

Stuff To Get

- A narrow glass container or jug, the taller the better
- 100ml (3½ fl oz) of each of the following:

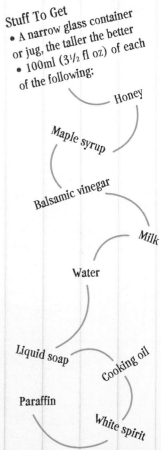

Honey

Maple syrup

Balsamic vinegar

Milk

Water

Liquid soap

Cooking oil

Paraffin

White spirit

Separation occurs!

The higher the density, the lower it will sink!

Try this at home – it's easy!

The Science

Different liquids have different densities and those with the highest densities will always sink below those with lower densities. In this trick the liquid with the highest density was added first, the one with the next highest density was added second, and so on. Density is measured by dividing mass by volume. For example, if 1 litre of water weighs 1kg, its density is equal to 1 as we saw above. The relative densities of the liquids used in this trick are as follows (in kg/litre):

Honey (1.3)   Liquid soap (0.93)
Maple syrup (1.25)   Cooking oil (0.9)
Balsamic vinegar (1.1)   Paraffin (0.8)
Milk (1.03)   White spirit (0.78)
Water (1)

Density is measured by dividing mass by volume

Wood-snapping Paper

PLEASE BE CAREFUL

If Godzilla had a fight with King Kong, who would win? It's the sort of conversation people (okay, boys) often have in the playground. Something else men might talk about is: can you use an ordinary boring newspaper to snap a wooden ruler in half? The answer to the second question is definitely yes, but the answer to the first is still unknown.

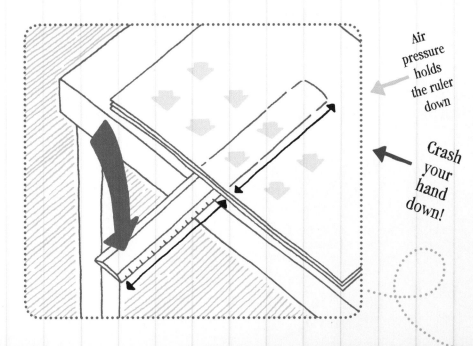

Air pressure holds the ruler down

Crash your hand down!

Let's Get Started

1. Put the ruler on the table, with about 40 per cent of it over the edge.

2. Place the sheet of newspaper on the table, covering the ruler right up to the edge of the table. Raise your hand; hold it there for some dramatic effect.

3. Crash your hand down on the part of the ruler overhanging the table. SNAP! It will break clean in half.

The Science

This trick works because of air pressure. The sheet of newspaper on top of the ruler is being pushed down by a large expanse of air. The air pressure holds the ruler down on the table enough to stop the ruler flipping the paper up when you hit the other end, allowing the force of your hand to break it in half.

Stuff To Get
- A wooden ruler (make sure it is okay for you to break it)
- One sheet of a large (broadsheet) newspaper
- A table

SNAP!

Did You Know?

For many years mathematicians asked the question: 'How can you draw a straight line?' Before you say, 'Duh, why not use a ruler?', remember these are mathematicians, and their line of enquiry originated from the fact that, while you could use a compass to draw a perfect circle, no such instrument existed to produce a straight line without reference guideways. The Peaucellier-Lipkin linkage, developed in 1864, managed to transform rotary motion into straight-line motion and was important for the development of the steam engine.

The Potato Skewer

Potatoes are quite hard, while straws are pretty weak and bendy, so it's a fair bet that it would be impossible to stick a straw into a raw potato. Think again ...

PLEASE BE CAREFUL

Stuff To Get
- A potato
- A plastic drinking straw

The Science

This trick works because, when you place your thumb over the end of the straw, the air molecules trapped inside become compressed and give the straw the extra strength it needs to cut through the skin and flesh of the potato – in effect, you are creating a solid tube. If the hole is left uncovered, the air simply passes through the straw, and it will crumple against the surface of the potato.

Did You Know?

Pressure is the amount of force acting per unit of area. The standard unit of measurement for pressure is the pascal (Pa). Measured in pascals, standard atmospheric pressure is 101,325 Pa, a shark bite can be 30 million Pa, steel can withstand a pressure of 40 million Pa and the pressure at the centre of the Earth could be as much as 400 billion Pa. Get a load of that!

Let's Get Started

1. Tell the kids that, if you can stick a straw into a potato without bending or breaking it, they have to do the washing up. They'll take the bet.

2. Hold the potato firmly in one hand with your fingers and thumb around the middle of the potato, not around the ends. Grip the straw firmly in the other hand, and making sure you cover the hole with your thumb, stab the straw sharply into the narrow end of the potato.

3. Go and watch TV while the kids wash up.

You must place your thumb over the top of the straw

Compressed air in straw

Take that, potato!

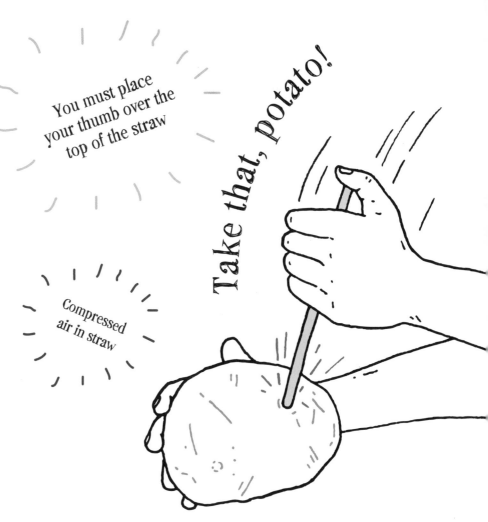

The Unpoppable Balloon

DANGER – ADULTS ONLY

When someone asks you to blow up a balloon they generally mean to inflate it, not blow it up ... KA-BOOM! But what happens if you inflate a balloon, and then try to blow it up by holding a flame underneath it? Let's have a look ...

Did You Know?

Three Frenchmen are credited with the invention of the first hot air balloon. In 1783, brothers Jacques and Joseph Montgolfier built a balloon for scientist Pilatre de Rozier, who launched a flight that lasted about ten minutes and covered 1.6km (1 mile). The passengers were a chicken, a duck and a sheep. Two months later the first manned flight took place and this time de Rozier was one of the passengers.

Let's Get Started

1. Start by showing what usually happens, so inflate the first balloon and tie the end to keep the air in. Light the candle and hold the balloon over it for a few seconds, and BANG! It explodes.

2. Ask your audience if they think it is possible to try again and for the balloon not to pop? They should say no.

3. This time pour some water into the ballloon, then inflate and tie it as before. Hold the balloon over the candle where the water is pooling, and ... nothing!

Stuff To Get
- Two balloons
- A candle
- Matches
- Water

Evaporation occurs!

Water inside the balloon absorbs heat

Hot air replaced by cooler air

The balloon will not pop!

The Science

The first balloon, your test balloon, bursts because the flame heats up the rubber and weakens it so it can no longer hold the pressure of the air inside the balloon. POP! With the second balloon, the water above the flame absorbs the heat through evaporation and begins to rise, and is then replaced by cooler water. This process keeps going, meaning that the balloon never pops. Literally cool, huh?

The Uncrushable Matchbox

FOR EVERYONE

A lot of bets are about strength. In this case, a matchbox appears flimsy and weak, so you would assume that anyone can crush it. But, like many things, first impressions can be deceptive.

Stuff To Get
- A matchbox
- A table or other flat surface

Let's Get Started

1. Make a bet with a friend that he cannot crush a matchbox, but, before he crashes his fist down onto the box, take it apart.

2. Stand the outer shell of the box on its end on a flat surface, and place the tray section, open side down, on top of it.

3. Watch while your friend fails miserably in his bid to crush the box.

The Science

The reconstructed matchbox is strong because the tray section disperses the force of a fist across its area, and the force is also directed along the length of the cardboard shell, giving the structure greater compressive strength.

CRUSH! Have a go!

The matchbox tray disperses the force of being crushed

Amazing compressive strength!

Did You Know?

Cardboard (strictly paperboard) boxes were first produced commercially in the UK in 1817, but the first pre-cut cardboard box was invented accidentally in the late 19th century by Robert Gair, an American printer and paper-bag manufacturer. One day, the ruler that was used for creasing bags accidentally cut the paper instead, and Gair realized he could cut and crease paper in one process to create boxes. And when Mr Kellogg invented his cornflakes in 1896, the market for cardboard boxes really took off.

Circling the Square

FOR EVERYONE

One of the tests given to small children to test their development is to ask them to draw a circle. Try it: it's not easy. However, this nifty bit of geometry will really help you out, and it's quite clever too.

Let's Get Started

1. Fold the piece of paper in half, and then in half again at the centre of the folded edge.

2. Next, fold the closed corner of the paper (in effect the centre of the piece of paper) in half, creating an arrow shape. Then fold the arrow shape in half again. Keep doing this as many times as you can.

3. Using the scissors, cut a straight line across the paper, about 3cm (just over 1in) from the point. Open it out and you will have a circular piece of paper – amazing!

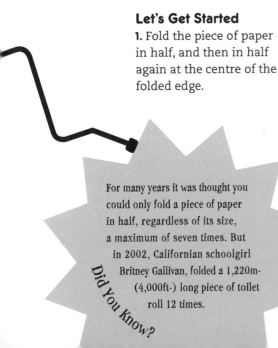

For many years it was thought you could only fold a piece of paper in half, regardless of its size, a maximum of seven times. But in 2002, Californian schoolgirl Britney Gallivan, folded a 1,220m- (4,000ft-) long piece of toilet roll 12 times.

Did You Know?

Stuff To Get
- A piece of A4 paper
- A pair of scissors

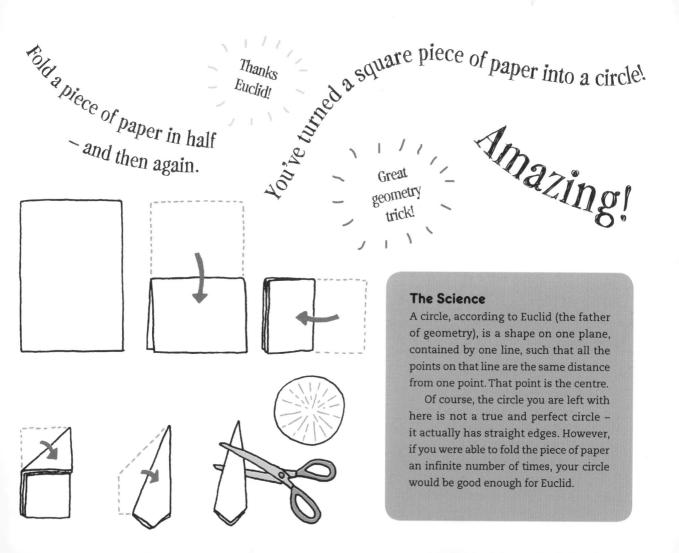

Fold a piece of paper in half – and then again.

Thanks Euclid!

You've turned a square piece of paper into a circle!

Great geometry trick!

Amazing!

The Science

A circle, according to Euclid (the father of geometry), is a shape on one plane, contained by one line, such that all the points on that line are the same distance from one point. That point is the centre.

Of course, the circle you are left with here is not a true and perfect circle – it actually has straight edges. However, if you were able to fold the piece of paper an infinite number of times, your circle would be good enough for Euclid.

Water Works

Parties are all about mixing and this trick should get everyone together by showing how water doesn't always mix, thanks to the magic of salt.

Let's Get Started

1. Fill two identical glasses with water right up to the brim.

2. Drip the food colouring and salt into one of the glasses and mix well. If you spill any water make sure you top it up to the brim again.

3. Now hold your piece of card over the top of the glass containing the pure water and, holding it tight, tip it upside down. Carefully place it on top of the other glass.

4. Slowly, making sure the glasses' rims are balanced on each other, pull out the piece of card – the water doesn't mix – amazing!

Stuff To Get
- Two identical glasses
- A piece of stiff card
- Five drops of food colouring
- 1 tbsp salt
- Water

Did You Know?
The Dead Sea (not actually a sea at all but a land-locked lake lying between Jordan and Israel) is 8.6 times saltier than the ocean, with a density of 1.24kg/litre. It is this that makes it easier for people to float in it, something that over a million visitors to the region do every year.

How does it do that?

The Science

The water containing the salt has a higher density than the pure water and will therefore stay at the bottom. Pure water has a density of 1kg/litre, while salt water has a density of 1.025kg/litre. To prove the point, start the experiment again, but this time put the salt and food colouring solution on top. Make sure you have a cloth standing by!

The water doesn't mix!

Salt water has a higher density

Show your friends!

A Hole in the Hand

FOR EVERYONE

A bird in the hand may well be worth two in the bush but how much is a hole in your hand worth? Probably not much but it won't hurt, you won't need surgery and you will definitely impress someone.

Stuff To Get
• A cardboard tube from a roll of kitchen paper
• A hand
• A view

The Science

Light enters both your eyes and these signals are translated in your brain into the things you see. Because the image from each eye is slightly different, the brain very cleverly makes sense of it for you so you do not see a double image.

In this trick both eyes are focused on the image at the end of the tube, the brain processes both these images into one and, hey presto, your friend has a hole in their hand.

Let's Get Started

1. Ask a friend to close their left eye. Give them the tube and tell them to hold it up to their right eye.

2. Now tell them to hold their left hand next to the far end of the tube, but warn them that you have made a hole in their hand – you wouldn't want them to scream.

3. Tell them to open their left eye. Cover your ears, because even though you warned them they might scream after all!

Double vision!

Look, a hole in the hand! Spooky.

Confuse your brain!

Light enters into your eye!

Your brain converts this light into signals

So Long Suckers

Sometimes a tricky task is described as being as difficult as pushing water up hill. The next time someone says that to you, just laugh and tell them you can do it – yes, you can make water travel upwards.

Stuff To Get
- A box of matches
- Eight small coins
- A plate
- A glass
- Water

Let's Get Started

1. Pour the water onto the plate and put the coins in a pile in the centre. Make sure the audience are watching the trick, not trying to take your money.

2. Light two or three matches all at once and lay them down, still burning, on top of the pile of coins.

3. Quickly put the glass over the matches. As the matches burn, the water will be sucked up into the glass.

The Science

This trick has caused a lot of arguments amongst amateur scientists. Many people think the water is sucked into the glass because the heat from the flame uses up the oxygen inside the glass and creates a vacuum, but this is not the case.

In fact, the heat of the flame increases the air pressure in the glass, forcing some air out. The flame goes out when all the oxygen is used up and the air cools and contracts. The air pressure outside the glass is now higher than inside and so water is sucked into the glass until the air pressure is equalized.

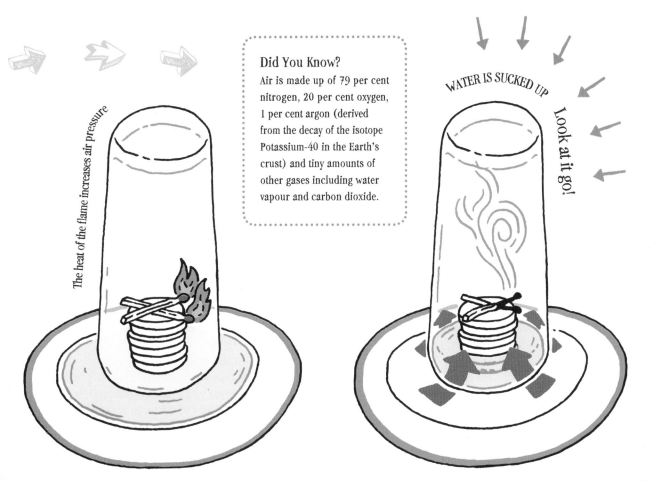

The heat of the flame increases air pressure

WATER IS SUCKED UP

Look at it go!

The Mint Fountain

PLEASE BE CAREFUL

When someone wins a Grand Prix they celebrate by wasting a whole bottle of champagne. This trick is cheaper, definitely more spectacular, and you don't even have to win a race to try it.

Did You Know?

Fritz Grobe, a professional juggler, and Stephen Voltz, a lawyer, created a giant cola fountain using 101 bottles of cola and 523 sweets, and created a video that went viral and received 800,000 views in its first week. The canny pair made sure they earned money for their explosive antics by sharing advertising revenue with their host website, and also reportedly held talks with the sweet manufacturers about a possible marketing deal. Now that really is cool science!

Let's Get Started

1. Carefully make a hole in the lid of the bottle using the skewer and do the same through the middle of the two mints.

2. Thread the string through the mints and the lid, so when you screw the lid back on the bottle, the mints are suspended inside the bottle but not touching the cola. You may need to tip out a little of the cola first so there is room for the mints.

3. Carry your bottle outside and place it well away from anyone or anything that you don't want to get covered in cola.

4. Pull the string up through the hole in the lid so the mints drop into the cola, and enjoy a massive coke fountain.

Stuff To Get

- Two chewy mint sweets (e.g. Mentos)
- A piece of thick string
- A metal skewer
- A 2 litre bottle of diet cola (not as sticky as the regular version!)

Carbon dioxide bubbles

KA-BOOM!

Minty explosion!

Stand well back!

The Science

The eruption produced in this trick is an exaggerated version of other experiments involving carbon dioxide. However, there is some debate over why it is that the chewy mint sweets cause such an extreme reaction. There is probably more than one factor at work here. First, the ingredients in the sweets that make them chewy (gelatine and gum Arabic) dissolve in the cola helping to break up the surface tension of the water and create more bubbles. Second, as a sweet begins to dissolve, its surface becomes covered in tiny pits that act as nucleation sites for carbon dioxide bubbles to form. And third, as the sweets are heavy, they sink to the bottom of the bottle, all the gas that is produced is pushed upwards, and ... KA-BOOM!

Bicarbonate of soda (baking soda)
You may well find this in the cupboard at home. If not it will be in the baking section of the supermarket.

Cornflour
Another store cupboard staple, also found in the baking section of the supermarket.

Effervescent pain relief tablets
e.g. Alka-Seltzer. Available from chemists or the pharmacy section of the supermarket.

Food colouring
Found in the baking section of the supermarket.

Paraffin
Available from hardware shops.

Petroleum jelly
e.g. Vaseline. Available from chemists or the pharmacy section of the supermarket.

Sodium acetate
Available online.

Surgical spirit
Available from chemists.

White spirit
Available from hardware shops (if you can't find any in the garage).

White vinegar
Called 'distilled malt vinegar' in supermarkets. Unlike ordinary malt vinegar, which is brown, white vinegar is clear.

Whitening toothpaste
Available from chemists or the pharmacy section of the supermarket.

WHERE YOU CAN BUY STUFF

Most of the items in this book can be found at home, probably in the back of a drawer somewhere in the kitchen, but there are some items that may require a trip to the shops.

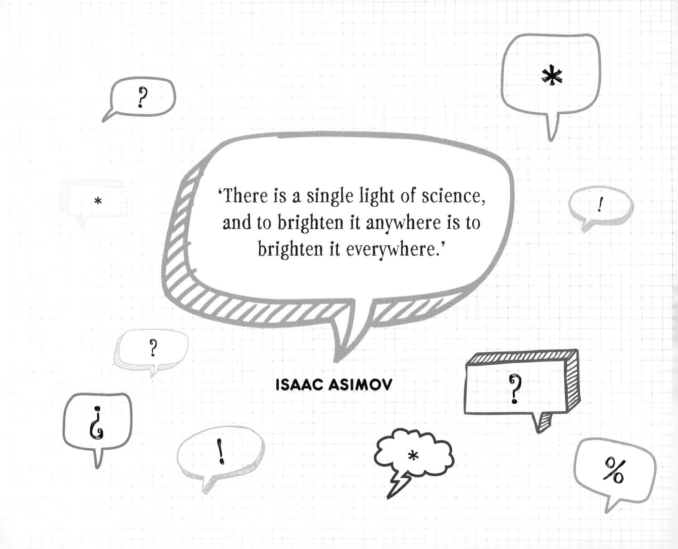

'There is a single light of science,
and to brighten it anywhere is to
brighten it everywhere.'

ISAAC ASIMOV